Dancing with Nature

by Peter Beamish

Sequel to
'Dancing With Whales'

Dancing with Nature

by Peter Beamish

Sequel to
'Dancing With Whales'

Trafford PUBLISHING®

Order this book online at www.trafford.com
or email orders@trafford.com

Most Trafford titles are also available at major online book retailers.

Front Cover: Photographs by Christine Beamish and © M.S. Bornstein (BEE)
Back Cover: Photographs by Nicholas Beamish

Printed in the United States of America.

ISBN: 978-1-4269-6305-6 (sc)
ISBN: 978-1-4269-6438-1 (e)

Trafford rev. 06/30/2011

 www.trafford.com

North America & International
toll-free: 1 888 232 4444 (USA & Canada)
phone: 250 383 6864 ♦ fax: 812 355 4082

To my wife Christine

and

our two sons

Nicholas and Timothy

CONTENTS

"DANCING WITH NATURE"

Preface

The 'House of Nature' has more than just a Darwinian 'Chamber of Evolution!' Some animals, indeed many organisms, may live portions of their lives in a separate part of such a 'House;' let us call it a 'Chamber of Altruism.' There also appears to be an intermediate state of Nature where communications and associated behaviours have characteristics of both the above perceived, profound 'Life Theories.' Thus, there are at least three 'chambers' in our biosphere. The rhythmic movement of living beings, from room to room within this metaphorical 'House,' may well represent a major component of a more genuine 'Music of Nature.'

Our family lives in a village inn. Upon entering centrally, one first encounters three unique rooms. On the far right is a sitting room, filled with signals: photographs, whale bones, notes of piano music and visions of video entertainment. Here also reside various signs and symbols, such as the caption on a child's drawing or a national flag. We could let this room represent the evolutionary environment of 'fight or flight' (both of these dynamic words being signals), and "survival of the fittest." Such a 'world' is where most humans spend much of their lives. These times, unfortunately, are associated with relatively high stress and a distinct lack of what author Paul Wilson calls, 'Deep Calm.' Even the ingenious Charles Darwin may have believed that evolutionary theory was correct, but not complete. Let us momentarily leave this chamber, which is dependent on what we call 'Signal Based Communication, SBC,' and move into a central chamber, a 'welcoming location' within the inn.

We now view, on the far left, a dining room representing a third 'Chamber of Nature.' How can we tell the differences in this area and why were such contrasts not noticed before? Answers lie in the recent discovery of another form of both the sending and the receiving of messages, now called 'Rhythm Based Communication, RBC.' In our metaphorical sense the rooms have different wall decorations, different ambience, and in the case of this left side 'Chamber of Altruism' all messages are encoded in 'Rhythm Based Time, T,' and not in signals, signs and symbols. Here, is found low stress and calm and in such an environment, we believe, 'The Great Whales' mainly reside. Perhaps only a relatively few humans have entered this 'world' but from a vantage point of the central chamber of Signal <u>AND</u> 'Rhythm Based Communication', we can now begin to understand the differences between two main 'Life Theories:' a) evolution and b) altruism.

In the sitting room (of 'SBC'), time is strictly conventional, moving in one direction from past to present to future. We are taught so in academics. But without the 'placing' of this linear time concept (and its associated behaviour) into an accessible mental 'safety deposit box,' and without storing it, temporarily, on a bookcase at the border between the sitting room and the central chamber, we will have a hard time leaving the 'Chamber of Evolution.' And it will be difficult both to live under the reduced stress of the central chamber, and to view into the main milieu of 'The Great Whales,' the world of mainly altruistic 'Deep Calm.'

In the dining room (of 'RBC'), 'Time' is of another form (and so, only to differentiate it, we use an upper case 'T'). It is called 'Rhythm Based Time;' its real part is always in the present, and it is compressible and expandable (sometimes we feel, even when in the central chamber, that 'Time flies' or 'Time drags'). Here is located a great magnitude and pattern of power developing from knowledge, as organisms display altruism, whereby they give beneficial information or perform helpful acts without any contemplation of reward. Here, every living being stands to gain and there are none of the higher stresses which are associated with exchanges of material goods, as found in the other chambers. Additionally, often, no organism using signals and evolutionary theory can by learning rhythms, outsmart and out survive the true altruist, as the higher stress of such a predator may be recognized, probably at a safe distance, by its lack of both synchronization and correct rhythms. Both organisms may then mentally and physically revert to the sitting room with its 'fight or flight' behaviour, which we document so vividly.

The simplest and the most powerful encryption system lives in the 'Chamber of Altruism.' It is a proposed foundation of the social structures, cultures and behaviours of our biosphere.

Please see Glossary (near end).

'Dancing With Nature'

They now say that poetry need not rhyme!
So here's a try that one hopes you will enjoy.
This book combines adventure with new knowledge,
The adventure story is complete, but new knowledge can surely
grow.
Seems a giant step onward from 'Dancing With Whales,'
However, there were many revisions to make it clearer,
In order to aim at credible worldwide interests.

Why should world peoples be so inclined?
An answer lies in 'Nature's Discovered Tools,'
An 'InformaTion' that can travel throughout all cells,
It carries a temporal concept not taught in schools.
Not known to Newton, Einstein or modern science,
It's a 'Time' as found in the empathy of love,
And in an 'Altruistic Nature' where needs are met.

Perception of lateness relative to a synchronization agreed,
Between active minds throughout this planet Earth,
Where, for life, a 'Kingdom of InformaTion' lies within.
So space and 'Real TIME' do now so greatly differ,
For if 'No Mind, Thus No Real TIME,' No 'NOW,'
But 'All Real TIME Is NOW,' the rest,
Are but scalar labels as pricetags, colourtags.

An important part of this work appears to be,
The new, unique and truly rational concept of 'NOW!'
And so I hope that all enjoy this novel, novel,
An adventure of fun, truth and new knowledge,
With 'Rhythmic Concepts,' that may help us all,
To discover a unified 'Kingdom' within our biosphere,
And you may see, 'The CareTaker' cares for you and me.

CHAPTER 1

The Arrival

NICHOLAS IS NINETEEN. He has recently and rapidly matured to gain responsibility, vastly improved knowledge and fortunately, good, common sense. More than any other excitement in his enlarging life he anticipates a coming expedition of research resolve and adventure tourism. The announced journey will depart from our home in Trinity, Newfoundland, and will take us up the east coast of this spectacular, fascinating island, to mysteries of coastal Labrador. It will be a human whale, human animal 'communications crusade.' The incentive to contact, to travel within, and to absorb the 'Nowness of Nature' will be occasioned by migrating whales, yes, by specific, well known, friendly humpback and fin whales.

"Dad, who else will be coming?" he inquires, as we load a precious cargo of underwater sound transmitters aboard the vessel *Ceres*, an especially designed and large inflatable, with a rigid hull, profuse power and substantial safety devices.

"We're six, all told, so that we can sleep on board or in three tents ashore. Elliott will be the senior; he brings a lifelong interest in Nature. Hans, a native of Austria and the most physically fit, has vast wilderness experience and biological knowledge, while George will do a photo documentary including, hopefully, the very important scientific identification shots of close animal encounters. And there will be Alex, a Russian, possibly one of the most brilliant persons alive on planet Earth!"

"It sounds like a 'six man,' valuable research experience!" exclaims Nick, intending to jestingly include himself as now newly mature! He will be invaluable for the whale communication research. "True, but we'll put ashore as dictated by weather and whale movements. Your mother, brother, students, friends and local mariners might join us unless we decide to camp at an isolated, abandoned 'ghost village' or in one of the giant, macabre, coastal caves! The expedition, will have to be a 'Now Time' experience, because we'll try to stay with whales; they are, in my opinion, perhaps the supreme 'Now Time' animals in our biosphere."

"Which whales do you want to follow?" asks Nick.

"Mainly Ida, Andrew, Cecile, Hubert, Meg and Scott, humpbacks that are here right now, but also Scratch, Oscar and companions, familiar fin whales that are also in Trinity Bay."

"I heard that you were with Ida for over three hours this afternoon, in Spaniard's Cove. Did you get definitive, rhythmic communications data?" questions Nick.

"We did. I'll present a summary at tonight's after dinner meeting, which hopefully will include our four new guests. Their flights should have arrived by now, from Russia, Western Canada, Australia and the U.S. We can expect that they'll have jet lag! Well now, *Ceres* looks ready for the night as soon as we can use our new haul off, to move her away from the dock."

The wind calms as the sun silently sets in the western sky. Nearby at The Village Inn, our Trinity headquarters, all await our guests arrival. Having long known these four distinguished visitors, happily I anticipate their coming company while carefully mixing an Inn specialty, our popular Manhattan cocktails of celebrated Canadian rye, Angostura bitters, sweet vermouth and a cherry with just enough cherry liquor to balance the bitters, and then --- the three secret ingredients!

In the Inn kitchen are a dozen, sizable, succulent lobsters and giant loaves of hot, homemade bread. Dody has prepared such a welcome Newfoundland-style banquet, complete with partridgeberry pie and 'figgy duff,' for over a quarter of a century! But this should be her 'coup de maitre,' especially with a cured, capelin caviar hors d'oeuvre, freshly, steamed mussel appetizers and a choice of cold California Sauvignon Blanc or frosty, feisty, local beer!

The van arrives! Alex, who exits first, initiates his rousing Russian bear hugs; the others follow suit with our family foursome, even recognizing many of the Inn staff. Students carry bags up the 'Inn Escalator.' This is actually our antique front staircase and where one explains that pressing on the lower, large, rounded newel post, delivers psychic energy to whisk one and all up a single flight to 'Chambre City,' where guests reside! Elliott has the 'OnTime' regal suite, George the 'LateTime' chamber, also named for the late Lady Lacely, the original owners wife. Hans has the 'OffTime,' honeymoon ensuite and Alex the 'EarlyTime,' quiet, thinking pad, all rooms being entitled, as I explain, for communication 'windows' of our science! After a fun dinner we casually adjourn to 'Whale Archive Lounge,' where questions are asked about the gallery of photographs, baleen plates of many species and the enormous, extraordinary ear structures, bones of the largest animals ever to inhabit the Earth. In the room centre is a double vertebra of an enormous bowhead whale, killed in Labrador about 1525.

"To unwind after such a long journey those cocktails were just what the doctor ordered," announces George.

"Well, a whale doctor did order them," jests Elliott; "and besides he says that he prefers his occupation because he gets less complaints from the patients!"

"I, for one, preferred the lobsters with the local 'Black Orse' beer, and the 'figgy duff' with the dark 'Screech' rum," comments Hans.

Students fill the back seats of the conference lounge, as a research and entertainment occasion is about to begin; included are Nicholas and his younger brother Timothy. Christine, our hostess and Innkeeper, sits with the honoured guests in lounging chairs near the slide and movie screen, where she inquires about the rigors of their flights.

"Let's start with an official welcome to Trinity, which may be the oldest permanent European settlement in the Americas, and also a welcome to The Village Inn, town center, and oldest whale research laboratory on the island. Let me introduce, using these print photographs, the potential stars of our intended adventure. They are six humpback whales that we've now studied for many years, including today, and nine fin whales that are, as I speak, also not far away. These images of Ida's back and tail patterns are means of identification. This very special whale is partially responsible for what I consider to be the most definitive scientific hypothesis of my forty year cetological career. She and companion whales indicated to us, three years ago, her way of differentiating between a statement and a corresponding question. Immediately thereafter we were shown, using their natural communication strategy, a difference between the affirmative, and the negative, that is between a 'yes' and a 'no' answer. We've subsequently learned that the same 'rhythmic scheme' is used by other species; it may indeed be universal! Today, as an example, we asked many questions about the east coast fishery, and here, as a handout, Tim has prepared multiple copies of a computer printout of both our questions and answers."

There follow excited sounds, then silence, except for the strident shuffling of papers.

"When will this group, I believe that you now say 'heard,' being mainly connected by sound, of humpbacks head north to Labrador?" asks Hans.

"My guess is two to four days but this is one question that we can't ask Ida on any day, because it's known that whales and all animals, other than humans, think little in the future. The 'Great Whales' are called 'Zen Masters' mainly because they live almost entirely in the present. Our questions must be 'NowTime' messages. As an aside, closed words are used to differentiate this new temporal format."

"How can you be sure that they don't think in the future?" inquires a student named Jason.

"We've asked many questions about their plans for the next day or week and we persistently don't get any answers whatsoever, almost as if they didn't understand the questions. In addition, animals don't have the same anatomy that gives humans the ability to plan create and produce for the future; we know this from brain research."

"After that short introduction, I'll give you a very brief summary of our human-animal communications research followed by our 'future thinking' plans for the expedition at hand. We can have a short Newfoundland visual arts 'Whale/Eagle/Fox Show' followed hopefully by some sound sleep!"

"As you must all know, Alex and his students discovered that animals not only encode messages in signals, like the barking of a dog for a message such as 'I'm hungry,' but that they, also encode information in 'Time,' a new temporal type for which we use an upper case 'T,' to differentiate the temporal concepts. In its most elementary form this is simply the rhythmic duration between the signals. Our Japanese friends discovered that such 'Time' is not perceived by organisms as being the same as Newton's or Einstein's time! The latter seems to go from past to present to future but the former seems always in the present! Measurements of both, however, must be made relative to a clock. Cyclical clocks can have window segments of 'Rhythm Based Time or RBT,' or just 'windows.' Four such 'window segments' which we presently use every day are: 'onTime, lateTime, offTime and earlyTime.' Here one could also, use closed words and an upper case 'T,' to differentiate temporal concepts. These are foundations of our communications research."

"Two questions: firstly, what exactly do you mean by 'window segments'?" asks Kirk, a student leader of bald eagle and land mammal communications research.

"RBT' is circular 'Time' like your watch. After turning a complete cycle, it then repeats itself, over and over. There are four 'Time' segments used in our communications experiments and they are generally each of equal duration and symmetrical. As an example, for our one minute rhythm, with whales, using a four second 'window duration,' the 'onTime' window segment lies from 58-02 seconds, centered on the twelve of your watch. For clockwise rotation the 'lateTime' window segment, or just the 'late' window, lies from 13-17 seconds, centered on the three of your watch, and so forth. 'Windows' allow for variations of signal timing. They are 'Time' segments once again using an upper case 'T' to differentiate the 'Rhythmic Time' concepts, that admit to normal biological inaccuracies."

"Then secondly, how is 'RBT' different from the time on my watch or in my head?" asks Kirk.

"Do you recall the age old philosophic debate between 'Becoming and Being?' Well conventional, irreversible, linear time is associated with becoming, planning, creating and producing, whereas circular 'RBT' is associated with being in 'NowTime,' using closed words for the very important, novel concept of 'NowTime' and solving present problems. Furthermore 'RBT' is reversible, like the pendulum on that mantel clock, and appears to be a main mechanism used in some animal communications. This should become clearer as soon as we start using our 'event space,' diagramed model."

"So here's the present status of our research. If we use two computer programs, one for sending and receiving messages and the other for measuring animal stress, we can search for enough universal concepts amongst many species, to convince our colleagues of the importance of 'RBT' in the peaceful coexistence and coalescence of humans with Nature. This is our collective goal."

"Thinking in the future, like only the human species can, we hope to study, near Trinity, the chosen humpbacks and fin whales, then leave within a day of the humpback departure, following them north, if necessary to Labrador, time permitting. We'll try to touch base with you student groups doing terrestrial mammal communications research near Notre Dame or White Bay so that we can cooperatively study your land mammal experiments. Timothy and Joanna, are in charge of shore logistics and data analysis. Tomorrow, and before leaving, we'll try to demonstrate this new 'Rhythm Based Communication, RBC,' the encoding of information into 'RBT,' with your bald eagle experiments Kirk and with your red fox families Joanna and with as many species of whales as possible. If time and conditions permit we'll also demonstrate universal concepts with snowshoe hares, black bears, moose and caribou although we should expect better results with these animals when we move north into wild natural wilderness. The bear community on Newfoundland's Great Northern Peninsula is truly untamed, unstressed and from past experiences, considerably cordial!

"Nick, may we have the slides turned on before you check to see who has a 'tormenting' thirst, a local expression for a nagging or punishing thirst? Please excuse or enjoy the odd expression of the 'Newfoundland Language,' such as 'tormenting' instead of nagging! We'll show only 50 of our best photographs, but each is chosen from about 20,000, so don't expect to see all of these spectacular behaviours by noon tomorrow!"

"The opening slides show surface feeding of fin, humpback and minke whales followed by sequences on electronic tagging, sound recording and rescuing entrapped baleen whales. The tags have allowed us to discover that humpback whales[1] don't sleep at night, except for about an hour just before dawn, when they appear to rest motionless, breathing slowly but normally and surfacing about four times an hour. Perhaps they meditate. Longer lasting tags show that some whales rest below the surface, apparently holding their breath considerably longer than normal. To do this, physiologists believe that they must slow their metabolic rate and perhaps circulate most of their heart's blood only to and from their brain."

"This photograph shows humpback barnacles[2] which make loud crackling, pulsed sounds, as measured from our studies of entrapped whales, but only when they are below seven to ten meters in depth. Our communications and behavioural studies have now revealed the distinct possibility that indeed these sound transmitters of the whales are not only symbiotic but beyond, a step above.

[1]   Humpback whales (Megaptera novaeangliae);
[2]   Humpback barnacles, (Coronula diadema)

Both barnacles and whales we now consider to be altruistic. This word, which you may hear again, has become our favourite in the English language!"

"What does the whale do for the barnacles then?" asks Joanna.

"Besides giving a safe habitat with no conventional predators, the whales carry their Coronula each year, to the 'finest restaurants,' from the tropical breeding grounds to the high latitude feeding grounds and return!"

"Do you have slides of the actual 'Rhythm Based Communication' experiments?" asks Alex.

"They're next! Synchronized minds think alike Alex! More on that thought later though." In the back of the room clever son Timothy whispers, "But fools seldom differ, Dad!"

"The photographs consistently show humpbacks 10-20 meters from our vessel *Ceres,* minke whales closer and the giant fin whales[3] virtually touching each other and our drifting vessel at the same time! The computer operator is high on the main mast when we are with humpbacks or minke whales, but further down at the main console for the fin 'whale contacts.' One engine is off and raised, the other is at slowest speed, as you can see from the almost nonexistent wake behind the *Ceres.*"

"When we are performing these experiments, which we call 'dancing with whales,' humpbacks prefer to be a body length or more away, possibly so that they can use their long flippers and energetic tails to slap the ocean surface for signals. Minke whales favour close, visual signals using underwater body language when they are near to *Ceres,* but

---

[3]   Fin whales (Balaenoptera physalus)

fin whales actually gently touch the vessel so that everyone on board can often feel the shudder of the inflation chambers! Here is a photograph of a humpback actually leaping clear of the water or breaching, at about 20 metres range, as a signal component of a message to us and perhaps also as a concurrent component of another message to a different animal. We have recorded 113 consecutive breaches from a single humpback, possibly meaning: I love you, I love you, 113 times!"

George then rapidly questions, " Can you predict, Peter, when such an event will occur?"

"Under certain circumstances the answer is a definite yes. Firstly, we must have our green light on the right of the computer screen, indicating low animal stress. Then we have to have successfully obtained a 'reciprocal greeting' in 'RBC, or Rhythm Based Communication,' and finally we hope to have an indication of 'Time' compression, which is recognized by the computer. This will be seen as a flashing red light on the left, below the screen, and we believe it indicates feelings of both energy and excitement. Breaches, for new whales, usually come on the final component of a three or four part 'Time' compressed message, and are predictable. For example, on a one minute rhythm, they're predictable by 15 seconds or more and to an accuracy of the 'communications window' or usually plus or minus two seconds, often less. They do breach at other times during 'RBC', which I'll explain in due course."

A new student asks, "What is 'Time' compression?"

"It's a major property that makes 'Rhythm Based Time' or 'T,' with its upper case T, different from

'Conventional time, or t'. It's a double signal in an 'RBT' window thereby shortening a message by one rhythmic cycle. Recall that this new temporal form is circular, like your watch, and not linear. An example of 'Time' compression would be a 'passkey' or greeting with a double signal in the 'offTime' window, followed by an 'onTime' signal. The breach becomes predictable in that it almost always happens in the final 'onTime' window. The computer operator counts down from five to zero and the breach landing most frequently occurs at or very near to the number zero. This is a photographers dream. We've even given folks enough time to change lenses and still get a good picture! The location of the breach must be estimated from behavioural context but the range is usually close to 20 meters for humpbacks. Let me show you a documentary videotape made recently by MGW, a major European film consortium, and you'll see the countdown in its vivid reality. You'll also see that fine underwater photographs can be made during these communication encounters, by simply holding the cameras over the side of *Ceres*. We do use a smaller stored inflatable, which can be launched in about a minute for photography as well as safety. This has also proved invaluable in many of our rescue operations when whales have been caught in near-shore nets. Skills emphasize the importance of rescuing both the whales and the nets; these two objectives are entirely compatible. For entrapped whales we have never had our computerized traffic light for animal stress change from even red to amber. So at these times, there seems no chance of 'Rhythm Based Communication, or RBC,' for which the light must be on green. It appears that highly stressed animals use 'Signal Based Communication,' and behave as Charles Darwin would have predicted."

"Time out," interjects Elliott. "Are you defining a class of communications specifically for evolutionary theory?"

"Precisely, but this is a substantial subject, a major theme for the expedition and a topic I would rather delay until tomorrow when most of your jet lag has synchronized to local time." The dimmer lights of the 'Archive Lounge and Theatre' are turned slowly from low to mid range."

"You are often using this concept of synchronization, and I'd certainly like a definition or some clarification," says George.

"Synchronization, perhaps the most important word in introductory 'RBC' means 'happening at the same time,' and in this case I refer to our guests' jet lags, their diurnal, or daily rhythms, adjusting to the same timing as ours. There's a short story I'd like to tell you before we adjourn; it may help to clarify this definition, and I hope that you find it as fascinating as I did. When we have taught humpback whales our communication concept of synchronization, if in pairs, they respond usually by exhaling together, instead of the more normal behaviour of surfacing one behind the other. In many instances now we have tested this communication discovery by repeating such conditional response experiments. Then recently, after 11 such repetitions, Ida and Andrew responded, with synchronized three-quarter breaches about 20 metres in front of the boat! All but the first were captured on camera. We were so astounded and pleased that Tim scanned a similar one of these remarkable photographs onto our web site for anyone, anywhere to be able to view.

Over there is an example picture on the fireplace mantle."

"A conceptual interpretation currently appears to be: 'We are bored with your repetitive conditioning lessons, now we will demonstrate synchronization in our own way.' This sight would have surely won a gold medal in the last Olympics for synchronized swimming! Why don't we adjourn and let our travel weary guests get some synchronized sleep? Would all students assemble after breakfast, s'il vous plait?"

Alex speaks up, "My friends at the Academy of Sciences sent me off with a supply of vodka to celebrate at such a time as this, including a flask with built in tumblers! Join me, Christine and shipmates! Let us put to good use Peter's suggested salutation of 'Chimo,' apparently from Canadian Inuit heritage, meaning 'To your great health and great happiness,' I've been told."

"And Peter, I did bring over a present from Mr. Yeltsin. It's upstairs and some of your students, if interested in holography, might like to see it. I'll use the escalator!" says Alex with a smile as he exits quietly.

"What took you to Sydney, George 'me son?' Do you remember that expression? It means my friend, in the Newfoundland language!"

"I recall, me son; I've now returned to every province and almost every culture in Canada! Business takes me to 'the four corners.' I expect to visit Alex in his Kremlin office next month," replies George.

Then what should appear but the most dazzling Russian holographic image depicting a myriad of gears and, as Alex rotates these geometric symbols a few degrees, the wheels spin in one direction or the other, just as does 'Rhythm Based Time!'

"Your technologists have modeled circular 'NowTime' Alex, and thank you, this will be a Village Inn treasure far into the future."

Students gaze in full scale fascination, while Elliott and George quietly depart for the escalator and much needed, sublime sleep!

"I'd like to take a group of three to Eagle Beach as early as possible tomorrow and stay at least one night; the weather report is excellent," remarks Kirk.

"Good idea! And then *Ceres* can stop to see your bald eagle and beaver experiments if opportunity permits. Ask the kitchen to load up your waterproof food cases early; you could leave right after the morning meeting."

As I tell Alex and Hans how very good it is to have them join this important scientific expedition, some of the students take their leave to local residences. Hans gently exclaims that he doesn't understand either Alex's original theoretical physics or the revised theory based on a new form of 'Time,' but I explain how important it is that he join us and assure him of the high probability of both a fine adventure and fun. He agrees. Hans and George are possibly the two most successful business entrepreneurs in the land and each has the very finest character for compatibility, useful for the coming expedition. Elliott and George command reputations that are more than 'household words;' their lives are astonishingly positive 'Canadian stories.' It is a distinct privilege to have so much profound intellect on such an adventure. One idea of this expedition is to match the boat crew brain power to the presumed intellect of Ida and Andrew, that seems to enhance definitive human whale communications!

"Alex, do you need some sleep?"

"Da! Oh, I mean yes, but first I must ask you a question to organize my thoughts during the night," comments Alex.

"Are we alike in waking early to resolve a conceptual problem and then falling asleep again when it's solved?"

"Yes, that we are," replies Alex.

"Well in such a case you should fall asleep again until you wake and hopefully you'll then be near local time with your biological clocks."

"I need an answer to this question. Why are you and the Americans so sure that whales and possibly other animals use a different form of time than our simple human 'perception of a clock,' a time that we all call Einstein's time?" asks Alex.

"This revelation came directly from experiment and not from any theoretical insight in the early hours of morning! If we commence tomorrow with clockwise, rhythmical computerized 'Time' and identify a noun such as 'Spaniard's Cove,' using 'RBC' then by conditional response, these particular humpbacks have taught us the following: if we are outside such a location and reverse the rhythmical 'Time' to counterclockwise then a simple noun becomes a corresponding question, such as 'Spaniard's Cove?' Interpretation based on whale behaviour is: 'Are you going to Spaniard's Cove?' This is experimentally verified because they have also taught us two other concepts, namely 'Yes' and 'No.' Their 'Time,' which in one direction encodes a statement, conversely in the other it becomes the corresponding question! Their 'Time' has two directions in contrast to ours, which only goes in one direction, from past to future."

"This I'll want to see," explains Alex. "Nevertheless I'll think on it tonight. Einstein was always baffled by a 'missing variable' to explain reality; I wonder if this could be it!"

"Some Australians call time 'the mystery of mysteries.' We similarly call it 'the enigma of enigmas.' That must be why it is such a fascinating challenge to study."

"If your data is true then we must find a way to combine the two types of time where the main condition will have to be that under no circumstances can the total time, let us call it upper case 'TIME,' be allowed to move backwards from the future into the past. That would change history, human behaviour and evolution and would not satisfy critical equations of physics," says Alex.

"You are perhaps the only man alive Alex, that can combine the theoretical physics and the necessary biology, to verify the conclusions to these experiments and formulate a new universal theory. I'll lay down all my cards, then we must try to make a grand slam in hearts, or better, a grand contribution to the very heart of human knowledge, about Nature that surrounds us."

Nicholas and Timothy are listening intently, but now, as if on cue from Alex's gentlemanly yawn, we all evaporate and escalate to the calm, composed, upper stories of The Village Inn.

Trinity Harbour

Trinity harbour the finest known,

For shape and size it stands alone.

The early explorers found the bay,

And many a gannet (1) led the way.

An Atlantic storm could make a sea,

But a ship inside would safely be.

Then onward sail to Labrador,

Red Bay, then Lisbon of distant shore.

And long before first white man roam,

The Beothucks could have an ideal home.

(1) Gannet *(Marus bassanus)*

CHAPTER 2

The Next Morning

The next morning Christine and I rise with the sun to a glorious, cloudless day. I check the computer mail while hot oatmeal and French toast are 'on the go' close by. The Columbian coffee aroma permeates the inner sanctums of my soul as I rapidly write ideas, mail replies, and formulate day plans. Then suddenly, solemnly and stately, in glides Alex, eager to speak!

"Good morning. The equations are satisfied; we do indeed have a realistic 'NowTIME', with such an upper case 'TIME' equal to 'Conventional time t' plus 'Rhythm Based Time, T or RBT,' two components for our theory," says Alex with a sparkle in both of his sleep filled eyes.

"It must be that your unconscious worked all night Alex. I'll take notes so that you can get back to bed as soon as possible."

"Notes nonsense! Here it is! - It must be that the essential condition for 'Rhythm Based Time, RBT,' is that it lies exactly at right angles, how do you say it, orthogonal, to the linear, timed 'Mental Vector Processes' of relativity."

"By that you must mean the 'Mental Vectors' which are carrying temporal scalar labels."

"Yes, exactly" replies Alex, and he continues, "physics and our reality of Earthbound events collapses if 'Conventional time, t' goes backwards, that is from the future to the past, as that would mean that an effect could produce a cause instead of the accepted fact that causes precede their related effects or results. But the concepts and equations of theoretical biophysics cannot survive without the modern concept of a two way rotation, like the Earth, as viewed from either pole, and additionally without your 'Event Space NowTIME' concept," Alex exclaims excitedly.

"Do you mean the concept that we all live in an 'Event Space,' and that time outside of such a space is in reality mass and/or energy movements with attached time, scalar labels, in fact, Einstein's 'reading-ON-a-clock,' time labels?"

"Yes indeed, and as this energy travels through your 'Event Space,' like that sound coming from your radio, then 'RBT' is orthogonal, similar to a spin on the sound waves," says Alex.

"Do sound waves spin?"

"Every particle apparently spins, from a quark, to electrons, to protons and as well to those masses that are a great deal larger. Biological organisms along with many measurable entities in physics, 'spin with rhythm' and thus possess independent 'Rhythm Based Times,' orthogonal to their individual energy movements through space. It is only the linear movements that are labeled with conventional time, or space divided by speed," adds Alex.

"So why the essential condition, the break-through of this morning?"

"Because both forms of time are labels on energy and/or mass vectors, that is, movements of energy or mass that must have both magnitude and 'spacial direction.' And when you add these vectors, the only situation where time cannot go backwards is for the circular labels or 'RBT' to be exactly at right angles to the linear labels, or to Einstein's time. That's the fascinating result of my calculations," says Alex, this time with considerably more excitement.

"Which allows 'Rhythm Based Time' to go in either direction, say clockwise or counterclockwise."

Alex then adds, "Proof that the two forms of time are indeed different!"

"Eureka!"

"We must try to solve additional aspects of the mystery of time during this expedition, Peter," says my Russian friend.

"And you must try now for some additional sleep, Alex," adds my wife Christine.

"A-OK, as you would say in your space program," and Alex solemnly, modestly and stately, glides back to his 'EarlyTime Suite', appropriately named, down the hallway a few meters.

Alex is tall with flowing white hair. He has great humility, great passion. great presence including dignity, charisma and charm. He manifests a magnetic personality, as close to a spiritual leader as one could imagine anyone amongst our eclectic, enigmatic species.

"Was it really worth getting all that excited?" asks Chris as she serves the steaming hot oatmeal.

"Knowledge is powerful; Alex has plenty. Whew! Lets just have a quiet breakfast."

Shortly later our student seminar commences in a second dining and meeting room.

"Attention please, we want to get this first meeting started, before any commencements of whale contact."

"Kirk's in the kitchen. I'll get him," says Sean, a university field assistant from England.

"While you're out could you ask Dody for toasted homemade bread, coffee and hot chocolate, please Sean."

"I'm back Doc, the cooks kept making me sample yesterday's baking!" says Kirk as he takes his place at the long, windowed table in our staff meeting and overflow dining area, adorned with Russ Hanson's amazing stroboscopic bird photographs, and more. This 'Eagle Room' has many paintings and photographs, some showing over a hundred bald eagles feeding on a capelin spawning beach just a few kilometers from Trinity. 'Perched' on a corner pedestal is 'Baldy' a beautifully carved wooden eagle, 'watching' over the varied activities of this research centre!

Kirk is a mature graduate student studying eagle ecology and communication at Stanford University in California. Since he had joined Ceta-Research three seasons earlier, he has graduated with honours from Harvard University and he has great promise to substantially enhance human-eagle communication throughout North America and perhaps further. Sean is Kirk's colleague in biophysics from last year and attends Cambridge University in England. This being his second season here, with eagles in Canada, and with Canadian beavers[4] he hopes to learn a lot more by working with Kirk, 'RBC' and Ceta-Research.

"As a preamble for this morning, an important event just happened. We've confirmation in both the mathematics and theoretical physics from Alex, that 'RBT' is indeed a new and separate temporal form. For the next meeting, therefore, I would like you to read a definitive paper on 'Event Space Spheres' and how they can be used to interpret the best model of our experimental data. Sandra will xerox copies for everyone as soon as she arrives. This morning I'll first give some definitions and then introduce our animal communications experimental techniques; these should be good for anyone, anywhere, anytime."

"We need computer printout on yesterdays fox data," says another field assistant while observing the overload light on her Palmtop, Psion computer.

"See me afterwards for all computer problems," adds Timothy.

"Right. Let's get started. 'RBC,' or 'Rhythm Based Communication,' the object of our joint research this summer, is defined as the encoding of messages in 'Rhythm Based Time,' or just 'Time,' remembering the upper case 'T.' This is something that we humans use occasionally. For an example consider 'love at first sight.' But generally we prefer to encode messages in signals, signs or symbols. Examples of the latter are words or visual signals, stop signs and flags. It can be shown that animals under low stress conditions, and using arbitrary but usually convenient signals, encode many

_____
4   Canadian beavers (Castor canadensis)

messages in 'RBT,' or 'Time,' and a possible reason why we didn't discover this earlier is revealed in the theoretical biophysics. This comprehension of a new communication, or 'RBC,' it would appear, uses a different temporal form than that to which we've all been accustomed. I'll say more on this aspect in a future meeting when you've had a chance to read and discuss the paper on 'Event Space Spheres,' including the theoretical characteristics of 'Rhythm Based Time.' A working definition of this new 'Time,' is given on the first page of the 'Lanzarote' reprint, which I think you should have all received by 'snail mail,' several weeks before your arrival. It is, and I quote: 'The Perception of Lateness Relative to OnTimeness.' The easiest key word here is 'Lateness.' Everyone knows how to be late, but please don't do it for these meetings! 'Perception' is with either the conscious mind, that is the central nervous system including the senses, or, the unconscious mind. And here we certainly get into complications of definition, relative to whichever authors you've studied. I'll give you my impressions of the unconscious as related to 'RBC' in another seminar. 'OnTimeness' is a new phrase roughly meaning synchronization but encompassing the concept of the initial point of an intended, shared or common rhythm between two or more organisms."

"Is the concept of 'OnTimeness' also defined experimentally, and exactly, by our computerized 'onTime' window?" asks Tim.

"Yes, definitely, and that perfectly introduces the important main topic for this morning of 'RBC' and General Experimental Technique, abstract number 19 in your handouts." (See also Appendix I.)

In comes Dody with plates of buttered toast and large pitchers of hot drinks; so I divert.

"After this meeting Tim, Joanna and Kirk will assign teams for red fox, caribou, snowshoe hare and eagle/beaver field 'RBC' experiments for today, with appropriate exchanges for tomorrow. Kirk will take three others to overnight at 'Eagle Beach' and *Ceres* can take up to four on each half day whale trip. Nick and Kirk can choose the four to join us this morning on the water. Everyone should get chances when 'whale contact' space is available."

Most of the student 'lunge feeding' (a term used in science, mainly with whales!) has subsided!

Holding up a sheet, I begin again. "This single piece of paper describes the experiments that we need you to do, it is a virtual printed recipe for 'RBC,' our newly discovered, human animal communications." (Appendix I.)

"Please see earlier references on experimental technique, for this unadorned but slightly specialized part of the student seminar. It can also be found on page 19, as mentioned above, in "A New Logic of Nature" and in an early abstract on 'www.animalcontact.com/research'."

"We need to talk about the elementary concept of a 'passkey,' the sequence of signals indicating one's desire to enter the world of 'Rhythm Based Communication'."

"Do I need a passkey to get back in, I'm off jogging?" comes a soft, inquisitive voice from the darkened, cocktail 'nook' nearby.

"Hans! You're an early riser for your 'jet set!' No keys necessary. Do you like flat or mountainous exercise, we've both?"

Hans is modest, generous and proportioned like a challenging marathon champion. He is a superb people person with so many fascinating tales of the very finest of family adventures.

"If you've both, I'll take both," he replies.

"The academic meeting is over. Does anyone want to jog up Gun Hill with Hans, hopefully be the first to count the humpbacks in the bay and then show the way to Taverner's Point?"

"I'll go," says Justin, who also prides himself on physical fitness.

Little do we know that Hans is in pristine condition and that he may well shatter the eight minute, long standing record from Inn to 'mountain top.'

Kirk and Joanna sort out logistics. Tim answers computer questions and Nick asks the chosen and excited students to help him haul in and prepare the *Ceres* for one of her most important expeditions. The coming voyage will carry four of the finest thinkers, critics, humourists, and potential promoters, in the multi decade history of this most resplendent laboratory for 'human-whale, contact experiments,' in Eastern Canada, in the continent and beyond.

Hans returns, having left Justin far behind! George is up, Elliott stirring. Being the Innkeeper's husband or, "the plumber," I check to see that the hot water circulating system is indeed allowing our no wait, "shower on/charge in," 21st century conveniences to be in peak performance. Others have said and I quote: "14 showers on a surface well - ha!" But it works, even if we're forced to get the fire truck to fill our pure and precious water source, during the occasional drought! At such a time Chris will put up artistic notes by basins saying: "Please conserve water," and I, often put up additional ones saying:

"Do not aspire to drink right here. We'd rather you try to sample the beer!"

Being a big breakfast fan and as guests gather in the "Victoria Room" for their first breakfast, I join them for my third, pancakes and fresh coffee! Presently, Alex arrives looking much improved and seeming a bit excited about the coming experiments. Nicholas joins us, complete with a computer demonstration and immediately George wants to know how the stress meter works.

"It isn't complicated," adds Nick. "We've a simple subroutine to the main translation algorithm or software program. At the beginning, and if a whale sends signals mainly outside the four 'RBC' 'windows,' then the red light comes on like this. Now I'll enter the times of various signals, say from any animal. If a statistical percentage of signals fall within the 'windows,' which are in this case, four second 'Time' slots, then an amber light comes on. If most of the signals are rhythmic, that is, they fall into the correct timing of the windows, then we've reached what we consider to be low stress in the animal and a green light comes on. We've adjusted the statistics whereby 'RBC' occurs during a green light and 'Signal Based Communication' is more associated with a red light. That's it!"

"Incredible! I could use such a stress meter during a board meeting!" adds George.

"What was your time to the top of Gun Hill Hans?"

"About seven and a half minutes," he replies. "Great place. The bay is full of whales."

Everyone looks up! There's an instant feeling of calm approval.

"Good morning, Penny. We've six here for ten o'clock on the water and we'll need some of Dody's good breakfasts to tide us over."

"You'll get tied if you don't hold your horses!" replies our most delightful, sarcastic, lovable waitress Penny!

"Speaking of tide, what's the high-low difference here and what's the present level?" asks Elliott.

Nick replies, "It's dead low now, great for beach combing; it should rise about two meters."

After breakfast and a short break, the ten of us meet for a debriefing on safety, guide guest communication skills for whale sightings, photography and other research, followed by natural history and logistics.

"The hull of the *Ceres* is of solid, foam fiberglass, designed to float 24 adults plus two engines. If we broke it in half then 12 could sit on the bow section upside down, use either of two radio systems and sing songs until rescued! Each of five independent floatation chambers will support 24 adults, making six boats so far and we have a seventh, stowed in the console, which can be inflated in under a minute using an electric pump, or under two minutes manually. There are ample life jackets where marked, 12 flares, two fire extinguishers and a first aid kit. The most dangerous part of the entire expedition is over, that's your drive from the St. John's airport!"

"We use five components of communication about whales. First is the warning word 'blow' which simply means that 'one sees a live whale, dolphin or porpoise on the surface somewhere.' Second, and needed quickly, is the 'clock bearing' where twelve o'clock is ahead and one hour equals 30 degrees. Most approaches are done slowly on our forward port side, at our bearing of ten to eleven o'clock, and the whale's aft right flank, at its four to five o'clock. Someone always sees an animal at three o'clock and shouts nine so that everyone looks in the exact wrong direction!"

"Third is the 'range' to the animal, in meters, which for photography is particularly valuable for focusing at short distances. If we reverse the order and give range first and bearing second then this is an estimate of where the animal is underwater. Fourth is the 'heading' of the whale as: 'left to right, into the bay or towards our boat.' This should help you determine where to look for the next surfacing."

"And last but most important are some of the multitude of 'comments' that can be useful with whale sightings. Examples today could be species, name, behaviour such as feeding, play etc., and the specific progress that we are making in 'Rhythm Based Communication.' In our baseball analogy, first base is 'possible synchronization,' second base is 'message mimicry,' third base is the 'communicaTion' of an interrogative 'Time' reversal, and fourth base is the transmitting of questions from an established vocabulary, including a home run of cooperative, altruistic answers, and thence meaningful communications. Don't fear, all this should become clear, with soon envisioned, live examples!"

"I'll ask Nick to do the natural history and logistics and we should meet in the back 'Mustang Room,' our place for getting your one piece exposure suits, in about 15 minutes or so. I'll be checking on various student field projects. We'll leave for expedition cruise number one in about a half hour, or as we say in this research business, T minus 30 and counting!"

CHAPTER 3

"Whale-Contact"

"T minus one minute and counting. Gentlemen please place your chair backs and tabletops in the upright position for takeoff! A-OK Nick, castaway. T minus zero. Up slow to half speed."

"What was that upright stuff?" asks Edward, a new British student.

"Just Dad's attempted humour Ed," says Nick, with a smile.

"Practice bearings everyone: one o'clock - Admiral's Island; four o'clock - ruins of an old whaling station; eight and eleven o'clock - church steeples; two o'clock - Fort Point lighthouse."

We are cruising at 30 knots on a glassy calm, spectacular harbour. A slow practice turn to starboard is announced and the *Ceres* leans gently into the curve like a Formula One racing car on a sharp, 90 degree, banked bend. Down speed to zero. Engines off. The 'sound of silence.'

"Welcome to 'Admiral's Island Airport,' home of about 100 pairs of Arctic terns[5]."

---

[5]   Arctic terns (Sterna paradisaea)

In a low voice, Nicholas describes their flight patterns and sounds, to Alex, Hans, Edward and Mark, all sitting astern. Mark is a young, bright, marine mammal student from New Zealand. Simultaneously I do the same for Elliott, George and two students sitting forward of the console.

"Notice the cannons at two o'clock, left by the British to protect the harbour mouth. There are others, tagged by our historical society, just ahead under water."

"Up slow to half speed." A moment passes while Nick and I search the horizon for whale blows. "The rock statue on the right was named the 'naked man' hundreds of years ago and when we tried to get it changed to the 'naked person,' in honour of all liberated women, many fishermen refused because there's a 'naked lady' on the other side of this point. It was aptly named the naked god, before the first European settlers."

We stop in sheltered waters.

"Bald eagle, two o'clock, on top of the largest pine tree," announces Nick. "It's an adult. You can tell from the snowy white on the tail and top of its head."

"We are in 'Green Island Tickle' and between eight and ten o'clock, on the island, live seven species of seabirds, but no eagles!"

"And a 'tickle' is?" asks Charles, a biophysics graduate from the University of British Columbia, in Vancouver, western Canada.

"A Newfoundland term for the body of water, generally between a peninsula and an island. It was named, presumably because that is where a boat could 'tickle' its bottom, or sides!"

"Blow, eleven o'clock, 100 meters, heading our way, minke whale[6], adult, feeding behaviour. Slow ahead on one engine."

"Eagle flying, three o'clock, another is chasing it, wing spans over two meters," declares Nick. "And do you notice how the primaries are bent upwards when they fly toward or away from us?"

"Where's the whale?" inquires George, with camera and large zoom lens at the ready.

"Estimated position is 50 metres at 11 o'clock. Aim your camera 30 degrees to the left of the boat's heading George, toward that column of gulls. Can you get the hydrophone ready Nick, we should try for corral feeding data?"

Ceres stops near hundreds of black legged kittiwakes[7] that form a vertical tower and as we record underwater acoustics the minke whale accelerates toward us producing a burst of mid range, audio, sound pulses stunning many hundreds of small fish named capelin[8]. The minke swoops under our bow as kittiwakes descend, competing for the dazed fish floating to the surface.

"The whale was feeding under the gulls, George, but it should surface very shortly, between nine and eleven o'clock."

"Blow," shouts Nicholas as about ten frames click off on George's power driven Nikon camera.

Shortly, we play the sound tape back at several speeds and sure enough the half second of recorded pulses seem like a machine gun barrage when played at a slower speed.

---

[6]   Minke whale (Balaenoptera acutorostrata)
[7]   Kittiwakes (Rissa trydactyla)
[8]   Capelin (Mallotus villosus)

"You have just witnessed what could be one of the finest examples of altruism in the animal kingdom."

"Please explain," remarks Elliott.

"We believe that the birds may have found the capelin for the whale, in cases like this, and then the whales killed enough fish for itself and for the kittiwakes."

"That sounds like 'cooperative feeding' to me," adds Hans.

"It's that and more Hans, because minke whales travel all over the bay and this one may never encounter these birds again. We have enough data now to suggest that these whales have a genuine genetic impulse to give, without any consideration of a future reward. Besides, whales have very little or no future thoughts, they live mainly in the present, they live in 'NowTime' and are masters at solving their own 'NowTime' problems."

"Hey Dad," Nicholas, whispers in my ear from the mast lookout position, "there are seven humpbacks blowing near the cliffs at Pigeon Cove."

Right thumb and forefinger close subtly together, into a 'circle signal,' responding confidentially to this good news.

"This minke probably won't come back now. Having had a reasonable meal we find that they usually go out into the bay for a short 'stroll.' We'll set our sights for a bigger non-feeding animal, so that we can hopefully commence communications research."

Up slow to three quarter speed, in along the spectacular shoreline of hills, harbours, anchorages, well marked fishing nets, water falls and eagles, we proceed, while finding the calmest waters.

"Harp seals[9] at eleven o'clock, just watch them dive together!"

About 50 seals all look at us until their leader communicates some sort of a synchronous dive concept. I wonder if it was a signal; one could expect that, with a high stress situation. Or was it a rhythmic prelude associated with the beat of a common biorhythm? One day soon humans may study both types of communication for many an organism throughout the biosphere.

"Range 5000 meters, bearing twelve o'clock, seven humpbacks in Spaniard's Cove," I shout.

"Assuming that Nicholas first saw the whales, from high up on the main mast," Hans comments to Alex, "good Lord, that young man must have fantastic sight."

"I concur," adds Alex. "I haven't seen whale backs or tails anywhere, excepting of course, for the minke."

"The way that Nick identified these whales at almost four miles was to watch for the low contrast blows against the dark cliffs behind. Polaroid glasses and a cap with a brim are a help. You'll find lots of such caps in the aft port locker, or you can hold your hand above your eyes to shield the glare and thus differentiate the blows."

Hans and Alex unpretentiously, don their sunglasses!

"It's Ida's group," shouts Nick from on high.

"Blow, and another at twelve o'clock, range 3000 meters, feeding along the capelin spawning beach at Pigeon Cove. More blows; do you see them Elliott?"

---

[9]   Harp seals (Pagophilus groenlandicus)

"That time I believe so," comes the reply.

"That time definitely," say together, George and Jay, the latter a second New Zealander, specializing in acoustic mammal behaviour

"Range a mile, bearing still straight ahead. Down slow to half speed."

"Sounding dive, large tail up - and - down. That's Andrew. I could see his large black dots on the underside of his right fluke," announces Nicholas.

"Come on down Nick, we'll start the computer program. We should use an 'alpha concept' of one minute as the whales are all in shallow water. Set your watches everyone. Starting with the next countdown we will be transmitting two second, underwater sound pulses every minute."

"Six whales are lunging after capelin along the shoreline and another is closer to us in deeper water," Nicholas broadcasts above the steady, moderate purring of the two, enormous, Honda, four stroke, 'super' engines.

"Down slow to dead slow," and the in-air sounds reduce to a low purring so that calm voices are all that's needed for the onboard discussions.

"5-4-3-2-1-mark!" shouts Nicholas. "First signal out."

"Starboard engine turned off, at the mark, Nick," as he had perceptively already started to enter such a signal as a comment on the computer data base.

Mark starts to move forward toward the bow where Jay, his travel companion, and Charles are now lying flat on the rather comfortable, inflatable chambers. "Why did Nick call my name?" he asks, while approaching the console!

"That's a timing command named after another Mark some years ago, but we're using it now so that we can have a count down to feed you to the whales, in a few moments!" quips Nick.

"Listen genius, you know that I couldn't fit down a humpback's throat!" counters Mark.

"5-4-3-2-1-markus!" repeats Nick. "Second signal gone."

"I'd rather Mark than Markus which sounds like Latin, Nickus!" retaliates Mark.

"OK you two, let's concentrate on the communications science."

With one engine off and the other in dead slow even the calmest talk is heard from the forward observers.

"The third signal is coming up, watch any or all of the seven whales for synchronization."

There follows another countdown to 'mark' and at that very moment we all see a huge blow on the closest whale.

"It's Andrew," declares Nick.

"Synchronization, one o'clock, 1,000 meters."

"Did you suspect that 'sync.' Dad, or just feel it?" asks Nick.

"The latter. What's the stress program reading?"

"Amber stress light folks; situation looking good," states Nick to all on board.

"Switch to the passkey program Nick and give a ten second warning before transmissions please."

Asking Alex and Hans to watch the coming manoeuvre, *Ceres* gently turns toward Andrew. Explanations about the next transmission are made, the 'first message signal,' which will be 90 seconds delayed in the 'offTime' window. Elliott and George glance at their watches.

"Hey, 'offTime,' that's what you called my room, one of your supposedly universal variables!" mentions Hans.

"Correct, it's actually our variable number three of ten, Hans."

"Warning for the first message signal, Dad," confides Nick.

"Thank you." Quickly, a second pinger, tethered to the console, is activated. It'll be tossed forward to hit the water precisely at the start of the main computer controlled transmission. Hans comes closer to observe.

"Down one stop on the main sound intensity; log that please."

"5-4-3-2-1-mark, first message signal, 'gamma one,' we call it," announces Nick. "It's lower in intensity by half the output power."

"Estimated location: 700 meters at twelve o'clock. He's probably resting on the bottom as it's only 150 meters deep where Andrew was last seen."

"Lunge feeding near the shore at two o'clock," reports Nick.

Capelin literally fly from the water surface just behind a lunging, hungry humpback and then a partner whale emerges with mouth unbelievably wide open and water rushing from the back sides of its mouth. Gulls hover over the second whale, which they must know is the main feeding one. They literally pluck the fish from the air without landing. The first animal, although probably just as hungry, has forced the fish into the second whale's mouth using its long white inside flipper. We then see these same two whales change places in a following lunge feeding incident in which case we have a primary example of mutually, 'cooperative feeding.'

"Andrew knows our 'passkey' so watch for synchronization on the coming, 'second message signal.' This one could be good for you to photograph, George."

"Warning on 'gamma two.' Dad," whispers Nick, as he has remounted to his mast position and he gently taps me with a boot tip, a practiced, concentration and focusing procedure.

"My dear Hans, will you take over on this new transmitter. It should go into the water on Nick's 'mark' but you should haul it out between two and three seconds later."

"A-OK as Alex would say! Hey! That's poetry!" replies Hans jubilantly.

"5-'flipper up'-4-3-2-and, down,' right on the mark for synchronization!" hollers Nick, with excitement in his voice. "We've green light, low stress conditions and we've a potential, 'reciprocal, overlapping greeting,' one of the terrific thrills of these new animal communication methods."

"The third message component will be 'onTime.' That's Elliott's room name or variable number one. Don't expect Andrew to signal 'onTime' but you can expect a signal in the 'offTime' window, 28-32 seconds after our transmission."

A ten second warning occurs, which Nicholas communicates via foot tap only! Hans seems ready. Nick announces that Andrew is between ten and eleven o'clock and then he shouts the countdown to 'mark.' This is the 'third message signal,' we call it 'alpha one.' Nick soon reaches down to point toward the expected bearing and then to steady George, by gently holding his shoulder.

"OK everyone, here comes Andrew's second signal. In 5 seconds -3-2- 'FULL BREACH'!!!"

Smash, just after the zero count, and as the 50 ton whale hits the water about half a second late, George is still clicking shots of the enormous splash! "I got the whole sequence! Great guide work Nick and thanks a lot for helping me to aim and keep steady," reflects George.

"We have finished our greeting or passkey. Next it'll be Andrew's turn to send his third signal which I predict will be a normal blow." (For actual photographs of such whale message signals, please refer to the 'Ceta-Research' section of the web pages at www.whalecontact.com). "Ten second warning everyone, 5-4-3-2-Blow, one second early," says Nick, (please see Appendix 1). By now we are nearly beside Andrew so *Ceres* goes to silent ship exactly on the next 'alpha time,' (by definition, the 'Rhythmic Time' centered in the 'onTime' window).

"Program for the interrogative of our rhythmic concept for fourth year capelin, Nick. Do you want me to look up the declarative in our dictionary?"

"I've got it here in memory, yes, it's: 'late-offTime-late-early.' So I just reverse the direction of our rhythmic time from clockwise to counterclockwise to get: 'early-offTime-early-late,' right maestro?" asks Nick.

"Yes indeed, carry on. But I'll turn down another stop on the sound transmission intensity if you'll log it, please."

It's amazing to see the sudden fascination in Alex and others at this reasonably routine guide conversation, which must be quite novel for them. This may now well be, in fact, a coming experimental demonstration of the quintessential communications discovery of our lives!

"How can you be sure that Andrew, the whale, both knows and remembers your concept for fourth year capelin?" questions Alex.

"Here we have Ida's 'heard' and they all know this concept. They learned it by conditional response when we always tested the capelin age, and then delivered the proper message, wherever they were feeding. They know many other rhythmic concepts as well, and they actually teach and frequently remind us to remember our own rhythmic concepts from past contacts."

"Ten second warning, Dad," says Nick and his computer conducts the music of the next four transmitted signals.

"Now watch for a double signal in either of the 'onTime' or 'offTime' windows."

"Review that again Peter, it is hard to keep track of all these signals," says Hans.

"The 'onTime' window, variable number one is now from 58 to 02 seconds on this master watch. The 'offTime' window, your room name Hans, is variable number three and is set from 28 to 32 seconds. It occurs every time, in this case, that the sweep second hand makes, a one minute revolution or one 'alpha rhythm.' Recall that an 'alpha rhythm,' measured in cycles of 'Rhythm Based Time' per unit of linear time is, in this case, one revolution per one minute. It is purposefully designed by us to exactly match the movement of the sweep second hand on your watch. Later, for eagles, the 'alpha rhythm' will be one cycle of 'Rhythm Based Time,' per twelve seconds. Let's watch for Andrew's answer; it can come now on any cycle" ('lateTime' is 13-17 seconds; 'earlyTime' is 43-47 seconds).

"Look up! 'Lob tail'- Smash, 'Lob tail' - Smash," Nick reports and records, while George photographs.

"A distinct double signal in the 'onTime' window meaning the affirmative!"

"And in simpler English doc?" asks Edward.

"We asked the whales if these capelin are four years old and they answered yes!"

"And in perhaps more philosophic terms?" asks Alex.

"We have demonstrated the interrogation of Nature under low stress conditions using what we consider to be a genuine communication system of Nature. For such cases we have no evidence of any replies other than with honesty and altruism."

"Well that certainly makes it different than language, wouldn't you agree Alex?" asks Edward.

"Ed, if I may call you that, deceit is mainly found in war!" replies Alex.

"Wouldn't you say that evolution with its, survival of the fittest, is a natural type of war?" questions Edward.

"Agree, but Peter's evidence so far shows that the honesty of altruism is a type of peace that is the opposite of war, just as 'true altruism' is so different from a business deal," responds Alex.

"Nick can you program the 'heard' size, how many, message please?" But Andrew is hungry and he is soon seen inshore with the other feeding humpbacks.

"Watch everyone, Andrew, our past communications leader is on a higher stress feeding break, but another should replace him and continue the identical rhythms and communications sequence."

"This, I've got to see," murmurs Elliott to George.

"Ten second warning Dad," says Nick again, and his computer conducts more music of the next four transmitted signals.

Then for Elliott, a miracle of miracles, as Ida emerges from the feeding whales and with a single synchronized tail slap and a 'lateTime' left flipper slap, answers the question with the number nine!

When I try to explain the answer, expected scepticism reigns! So we go to the very edge of the nearby beach at Pigeon Cove for a well deserved coffee break and discussion. Pigeon is a local term for sea pigeon or black guillemot[10], and as we approach the Cove, hundreds either dive (as they are excellent swimmers) or fly along the coast, where at other locations they can possibly capture a small fish for their dinner. When we arrive capelin are massing on both sides of the *Ceres* and Nicholas is quick to point out the actual spawning events.

"If you are observant, you can both see, and hear, from the boat's bow, three fish swimming directly up the beach with symmetric tail movements. They comprise a male on each side of a female and it is thought that the lower bodies of the males are actually helping to beat the eggs from the female onto the tiny pebbles of the shoreline," explains Nick.

"Coffee, tea or me!" I quietly say to the few who are unable to fit onto the bow of *Ceres*, at present.

---

[10]   Black guillemot (Cepphus grylle)

"Black coffee, no sugar," says Alex, "but we need you to explain Ida's behaviour."

Nicholas has now retrieved enough capelin using a casting net and he demonstrates by saying, "This is a male, slightly larger and with spawning ridges down both sides. This is a female, much more colourful, and as you can see, full of eggs. Would anyone like a picture? I'll align them, male, female and male, so that you can verify, how the ridges aid the process by 'locking' above the female so that the three symmetric tail motions better propel the fish to the highest possible beach position and also aid in forcing the eggs into the beach."

"If you folks up forward would like coffee, tea, cake, or cookies, then come on aft and perhaps every one can have the closer bow observation."

All four students jump at the thought of food and we move the restaurant to the stern of *Ceres*! Nicholas and Elliott have a fascinating talk about the behaviour of capelin and their predators, as George snaps not only fish pictures but people portraits, which hopefully will recall these precious moments for years to come. Many bald eagles[11] glare upon us from the steep, amphitheater like, cliffs above! They call to one another. Echoes reflect, again softer, and again still softer. The rocks are sedimentary, layered sandstones and limestones, and amongst the oldest of such known on the planet. In the centre of Pigeon Cove is an enormous, exposed mountain-like bending of the layers into what geologists call an anticline. I explain that one would possibly drill for oil on the margins of this feature if it was buried amongst

---

[11] Bald eagles (Haliaeetus leucocephalus)

other oil producing sediments. The stunning blue of wild iris line a central valley where bears descend for fishing. Entire ecosystems flourish where water seeps through the sedimentary layers and even the driest rocks are alive with colourful lichens where a bright orange seems to flavor the many favourite roosting spots of bald eagles.

"I can't drink either tea or coffee," says Mark mournfully, but with his delightful accent.

"Chris or Dody must have heard about that, you lucky devil, as here's some hot chocolate!"

"Unbelievable! Thanks," says Mark, grinning from ear to ear.

"Why not the coffee?"

"In three minutes I'd have to go ashore," replies the lad with slight embarrassment.

"You must have dozed in the 'logistics' orientation. We have a fine, easy to use, marine head under that stern seat, which even boasts a sailcloth surround and a great view!"

"Nicholas and Peter you promised to explain Ida's behaviour. How about up forward with everyone, as the significance is paramount?" Alex speaks loudly from the very bow.

"A-OK, 'Dr. Inspiration!' Nicholas will you move the *Ceres* off the beach; use just one engine, for extra safety, if you can?"

As we drift in deeper waters, four enormous bald eagles drop to the beach while hungry gulls wait on surrounding lower rocks. It isn't difficult to tell who probably owned the beach, long before humans entered onto such a stage!

"Ida, while ending her meal, must've received a message expressing Andrew's hunger. We believe that during our 'Rhythm Based Communication' encounters, all nearby whales are listening but only one is a whale-to-human communications leader. When Ida arrived near the *Ceres*, she took over that position and answered our often asked question with a tail slap followed by a left flipper slap. We have previously taught these animals, by conditional response, that a tail slap represents the number ten and a left flipper slap represents a minus one. So the answer is nine. For your interest in our instructed counting system, a right flipper slap represents a plus one."

"But one tail slap is then the signal ten, not a rhythm," perceptively states Alex.

"Exactly, you have seen how signals and rhythms can be juxtaposed in human-animal communications. Synchronized signals, as you saw, can contain both 'Signal Based Information' and 'Rhythm Based InformaTion'." (Please note the important use of the upper case "T" in the latter.)

"Does this example mean that whales can count?" asks Mark.

"All animals, we believe, can probably perceive quantity, but whether in their natural environment they use base ten, two or twelve, we don't yet know. Let me remind everyone that so far we have only studied human-animal communications using 'RBC,' but that we are still a fair ways away from studying whale-whale or animal-animal messages. The key difference is that in the former we create the rhythm base, but in the latter there most probably exists one or more rhythms, more natural to the organisms involved. The logic of this last statement is that various confidential combinations of rhythmic bases could make messages private, an important advantage in evolution. When the day comes that organisms share with us their confidential rhythms, humans will have finally and truly joined into the innermost nature of Nature."

Slowly Nicholas brings the boat further into Spaniard's Cove where the seven whales are engaged in cooperative feeding. A lead humpback forces the fish into the mouth of a trailing companion, using its very long white flippers for which it has been given the name *Megaptera*, meaning: 'giant wings,' in Greek. This is my favourite word in all languages.

"I understood that you now believe there are nine whales in this group," says Charles.

"I do. Look for the missing two. Besides, we now use the word 'heard,' for whales Charles, spelt 'h-e-a-r-d,' which is a pun on the verb to hear! This is because they're connected more by sound than by sight. The latter, that is vision, is the dominant sense for a 'h-e-r-d' of land mammals. The other humpbacks are probably a few miles along the shore, busy with their own cooperative feeding."

"Could we switch places so that you can program the: 'Cod fish, how many?' message, please Nick?"

However, on a far beach, just visible with binoculars, is one of Nature's surprising treasures!

CHAPTER 4

"Eagle Beach"

"**T**hat's a black bear[12] eating a salmon, just like an ear of corn on the cob, sitting back on its haunches looking like a queen or a king of the world! And three bald eagles are circling overhead!"

We slowly turn the *Ceres* toward the bear as George asks for help in steadying himself high on a seat directly in front of the console, a seat we call 'the couch.' Soon we are near the beach, without detection.

The last engine is turned off. "Silent ship everyone, except for George's clicks. Move like a hungry 'First Nation' hunter. Mark and Ed each man a paddle."

"Ten second warning Dad," whispers Nick as his computer is again about to conduct the underwater, inaudible music of three rhythmically contrasting signals.

We sit, now motionless, not a stone's throw from this adult, seemingly friendly, large black

---

[12]   Black bear (Ersus americanus)

bear, while I study the barely (no pun!) noticeable notches in its ears.

"That's, Joe! He must have taken that salmon from Lloyd's gill net, which is now on our starboard side, just here beside the boat. He's essentially finished that entire fish, leaving little for those eagles! Look at the masses of salmon, silver flashes underwater, still caught in the net. We'll use this 'sophisticated device,' Tim's bicycle bell, to start signals every 20 seconds!"

Joe, our black bear, twitches his ear in time with the third bell sound - 'Possible Synchronization.'

"Now, add the sound of this wrench being lightly tapped on the mast, Mark, as I count down for the same passkey that we used for the whales, 'offTime, offTime, onTime'. "

Joe mimics, as he has done before, indicating 'Successful SynchronizaTion.' His stress seems to lower and he wanders toward the net, toward us!

"Message complete to the humpbacks," announces Nick. "Answer as early as 30 seconds George, if you want to swing your camera to about six o'clock."

Our slightly raised voices now have no affect on Joe's behaviour. I study our computerized, bear dictionary to find out his mate's rhythmic name.

"Whale 'alpha time' in ten seconds everyone. 5-4-3-2-'SPYHOP!' That's Hubert, half out of the water and probably looking at us!" excitedly announces Nick.

"Add one soft hand clap starting at the mark, for the next message to Joe everyone, ready 5-4-3-2-1-mark-1-2."

The high pitch bell, the medium pitch mast sound and the low pitch clap all go out together. Joe seems to lower his stress again sitting like a curious student in this outdoor, congenial classroom! A second computer is now conducting the music of three sequences from ten persons, played in rhythmic contact to one, observant, adult, black bear. Joe replies in the 'onTime' window, pointing with his nose up a grassy hill. And there, at the top, is his mate whom we had previously named Maureen.

Like a curtain on a Shakespearian play which suddenly and dramatically drops down after the first scene, Joe moves quickly up the hill toward Maureen, and the whales move to deeper waters to digest their brunch. With feelings of serendipity therein, *Ceres* slowly turns along the spectacular shoreline, toward 'Eagle Beach.'

"Why do you think Hubert's spyhop was at the 'alpha time' or in the 'onTime' window?" asks Alex.

"To signify a 'Rhythm Based' or truthful answer to the question about fish biomass."

"And please explain that truthful answer," pressures Hans.

All are listening. Most are not yet ready to believe that the humpbacks did indeed answer this practical and important question, this very question that could soon improve the Northwest Atlantic and other world fisheries.

"Here's a case, well documented if George's pictures come out OK, where signals and rhythms, both previously learned from us, announced that Spaniard's cove was half full of cod, or as Newfoundlanders would say, half full of 'fish.' You can see them on the sounder right now, all the way from the bottom up to mid water."

"Then how did you know that the answer was half?" asks Elliott.

"Because Hubert rose vertically, approximately half way out of the water. Humpbacks, in fact, taught us that signal."

"How do you know that it was truthful?" questions Jay.

"Because: a, it happened in a rhythmic window of time and, b, all verified rhythmic answers to our 'RBC' questions have always been correct. These animals and others, especially when in rhythmic communication are altruistic. That's a priority, biological discovery of this research."

"Do you mean to say that animals move into the domain of evolutionary behaviour when they are under high biological stress and may then move into altruistic and truthful behaviour when that stress is lowered?" asks Alex.

"Yes precisely, and well stated."

"Richard Dawkins will debate with that," continues Alex.

"He is largely correct. Evolution is correct. I am an evolutionist like Richard. But evolution is not the complete scheme for all life! There is another 'chamber' in the 'House of Nature' and that room is a real, between-meal, low stress 'Chamber of Altruism.' All organisms move within the realm of evolutionary tendencies, trying to feed and solve stressful problems, meanwhile maintaining high internal tension. And then, whenever possible, they move to the realm of altruistic customs where, because of lower stress, they feel an inner contentment and its reward of a potentially healthier and more powerful mind, a greater 'survival potential,' and from both internal and external conditions, a potentially longer life."

Everyone relaxes as we slide along such scenic shores with eagles' nests and fishing nets. A sweeping, swerving flock of ruddy turnstones cross ahead and with all hands now forward, Hans observes, "I wonder how they navigate because I've read that no signal has been found for birds such as these, or fish, at the moment when they change direction."

"You're absolutely correct Hans, and if they can't make sharp turns with signals, like a regiment of soldiers, the only alternative is to encode the messages in 'Time' or, in other words, to use 'Rhythm Based CommunicaTion'!"

"How could that possibly work?" asks Mark, the one who, along with Alex, most often questions ideas!

"Elementary, 'me son,' the birds that you just saw have synchronous wing rhythm as well as other simultaneous biorhythms. A navigation leader then signals, to the flock, several units of a 'duraTion,' a duration in 'Time T,' NOT linear time t, before an 'onTime' window. Every bird learns from experience to turn a certain number of units of direction, and to do it on a beat. There must therefore be a known relationship between these signaled units of 'Time,' which we call an 'early concept,' and the units of directional change which could be, say, in degrees."

"OK, mister philosopher, then how do they know to turn left or right?" rapidly replies Mark.

"That's actually one of the best questions of the day! There must be a species evolved gene, which rules for signals in this case, and not for rhythms. As an example, if the leader is 'late' say of an 'onTime' window, then everyone turns left and, if early, they turn right. A more easily understood illustration of this idea with fish, is that if the leader is 'late' on a left tail motion, turn left; if 'late' on the right, turn right. The reason that this is an important question is that the answer, as described, combines both 'Signal Based Communication, SBC,' and 'Rhythm Based CommunicaTion, RBC,' in the same message. The two are entirely compatible!"

"Blow, 12 o'clock, 2 miles out in deeper water. This is a male sperm whale[13], probably surfacing after a long dive." There is much evidence on board of excitement, as probably most have never seen this famous species (like Moby Dick)!

"Blow on another sperm whale at 11:30, a mile beyond the first larger one. Can everyone see the blows being forward at about 45 degrees? Such is provided the whale is heading across our path, as in this case. The blows will appear vertical if we are in front or behind such a whale. This is a diagnostic feature of these, the largest of the toothed whales."

"What's a diagnostic feature Doc?" asks Mark.

"That's a characteristic found on no other species, Mark"

"Blow 1 o'clock on a third sperm whale," announces Nicholas, as I speed up in these calm waters.

"They are all large animals, easily over 60 feet long!"

"Blow number 3 on the first and closest whale. Two more exhalations and we could be almost

---

[13]   Sperm whale (Physeter catodon macrocephalus)

alongside. These seem all large males as we very rarely see the females of this polygamous species who mainly remain in tropical, warm waters with the calves."

"What do they feed on?" asks Hans.

"Benthic species like lumpfish, octopus, and giant squid."

"Blow number 4 on the nearest and largest animal" as I evenly slow continuously from this point to hopefully a near proximity of a friendly animal.

"Now we can see the single blowhole on the forward left side and, as well, the rounded, low dorsal fin."

Ceres slowly moves nearer the 60 plus foot sperm whale being over twice the length of our vessel. As usual this remains to appear 'RBC's low stress' and thus seemingly very safe. The grapefruit sized eye is now visible as are the peculiar wrinkles on the surfaced back of this animal, that we could call 'Physety.'

Having timed the previous blows, Nicholas is able to give one of his famous 'count-downs' for the coming close exhalation. George is at the ready, having taken many approach photographs.

"Ten seconds everyone" Nicholas is able to communicate to all as we are now almost on silent ship. "5-4-3-2-1-0-1-2-" and then an 'explosion' happens with hundreds of liters of air being ejected forward ahead of *Ceres*.

"Surface dive. 'Physety' should be up within a minute" as he turns underwater and peers at us with his other eye. We turn the *Ceres,* a trick learned for such a situation, with the port engine slow in reverse and the other slow ahead, and with the helm turned hard to port.

Using our sperm whale rhythmic code and an acoustic transmission at an extreme low volume we are able to ask the animal the only simple question for this species, developed by 'RBC.' This is: 'How many more surface dives before a deep 'sounding dive?' The answer came back a surprising zero!

So *Ceres* falls back behind 'Physety' for fluke photographs and Nick and I even prepare our own pocket cameras. Warning George, we once again assist him onto the 'console couch' where Nick can steady him in case of sudden movements.

"How much time Nick"

"About a minute Dad"

"Look for identification scars and shapes as the flukes should soon rise tens of feet in what has been called a 'sounding dive'."

Nick checks the master watch and estimates about 10 seconds which he broadcasts to all camera bearers, or in other words, everyone on board!

"5-4-3-2-1-sounding dive" 'Physety's' back arcs and the tail goes straight up, yes with notches on the left side where either another whale or a giant squid has taken a bite, possibly for a meal or such!

Alex from the bow is looking with his mouth open in amazement. Expressions as: 'Wow,' 'Amazing,' 'Unbelievable,' 'A-1,' and more, emanate from all over *Ceres*, but mainly from the bow which was nearest to the enormous tail! Then, all is quiet. We turn away toward shore. Slowly at first, we then speed towards the same beach where we studied Joe the salmon feeding black bear. Next we turn westward and round 'Rag Rocks' with a 'zillion' well fed gulls all facing into the mild breeze. There ahead in sheltered water are Kirk and assistants, anchored 50 meters from a notably large, roosting bald eagle!

"Silent ship everyone. Break out our paddles. *Ceres* to Kirk, how do you read, over."

"Loud, clear, we're nearing third base with Albert and we're asking the 'your nest, where?' question, over," rings out Kirk on our radio, so that all can hear.

Suddenly Alex, becoming inspiringly attentive, exclaims, "What's third base again?"

"Whisper everyone. Albert is a valuable eagle; we don't want to stress him during such an experiment. Third base in 'RBC' is 'transmitting questions.' You may recall that it follows 'synchronizaTion' and 'message mimicry.' Slow strokes with the paddles and together chaps, so that George can stand steady on the 'console couch,' with Nick holding him. What signals Kirk? We are 'silent ship,' over."

"Stand by one," comes the reply.

"Here are extra binoculars anyone, watch for body movements, and listen carefully."

"Kirk to *Ceres,* Albert is using alternating head motions and low pitch calls; transmission complete, over."

We watch in amazement as Albert lifts off in a northeast direction, circles and returns to his roosting perch. He then repeats this short flight while making a series of screeches which the eagle team is hopefully recording.

"I've got all of those sounds on our computer's eagle program, Dad," comments my amazing mate and research master, as if somehow reading my mind!

"*Ceres* to Kirk. Try the concept 'Do you have one chick?' if you wish, over."

"Roger."

"How does the eagle understand this?" asks Ed.

"By previous conditional response experiments where, in this case, we combine the rhythmic, learned, declarative concept for chick with the unity concept, and then reverse the direction of the 'Timing' sequences to make the interrogative massage."

As we paddle within 100 meters of Albert we can easily monitor Kirk's whistles, and the eagle replies. The first answer is in the negative, based on two rapid head motions six seconds late, in the 'offTime' window. With hand signals I suggest Kirk repeats the question with the number two replacing the one. He acknowledges agreement with the fore-finger thumb, circle signal.

"Watch for a possible reply to the question 'Do you have two chicks?' which is going out now," I whisper to everyone aboard *Ceres*.

The second reply is affirmative because we recorded two rapid low pitch calls in the 'onTime' window. Suddenly, thousands of fish hurl themselves onto the beach beside us! Albert swoops down, looks around and then feasts leisurely on only the more tender back muscles of the capelin. Included on his menu is the orange caviar, but not included at the restaurant, are hundreds of hungry gulls waiting 70-90 meters from the spawning fish. Two other eagles land a dozen meters from Albert; both are females. A few moments pass; the gull noise crescendos. As we slowly approach the spawning capelin, all of Albert's enormous wingspan majestically unfolds, close before our eyes, as he lifts dramatically off the beach and flies directly over our heads. He is then followed by the female eagles to his grandiose, grassy perch. Immediately, hundreds of noisy gulls swarm the beach. Their feeding action, mostly successful, lasts several minutes, until *Ceres* slowly lands amongst the myriad of flashing fish. The gulls that leave the beach with a capelin in their beak must defend themselves against attack after attack until the lucky ones find free air space in which to swallow! Almost immediately they then perform a 'shiver shudder,' a humorous gull's airborne shimmy, presumably to aid with their ingestion!

"All ashore for beaver break and beach snacks!"

Nicholas answers questions about the active spawning occurring near the boat; then he leads the brave ones to sample the very caviar that Albert relished!

"Never eat anything until your guide eats it first, my Dad would say," articulates Nick.

Soon he climbs aboard and we switch places after which he hands over equipment and staples. He will moor the *Ceres*, inflate the tender and join us, all in a few moments. Meanwhile further down the beach where no fish are spawning but where rocks make an ideal place to tie a smaller boat, Kirk has landed with his three assistants Justin, Jessica and Sean. These three are audibly vocal indicating a youthful sense of excitement. Justin and Jessica are returning to 'Eagle Beach' where they had both previously been on an earlier expedition, several years before. Some of our team slowly explore, some just stand in amazement at the sheer beauty, the now tranquil beach, the imposing sandstone cliffs, the splendid stands of spruce. And there, not far away, are the two missing humpbacks, in the company of minke whales, dining at their own 'Eagle Beach' marine mammal restaurant!

We meet on 'Moose Meadow,' near Kirk's well established summer research camp where gear is being stowed and a barbecue lit.

"I just heard it said by a student, 'this is my favourite place on planet Earth.' I'm beginning to understand why," says Elliott.

"Well I've seen hundreds that might qualify," replies Hans. "But this is certainly at or very near the top, except for one thing."

"And what might that be, it's isolation?" questions Alex.

"Oh no! That's a positive aspect, but on the negative side, I've seen evidence of black bears which, in my experience, can cause camping problems," replies Hans.

Kirk jumps in, "Bears are really no problem here in Newfoundland; under any stress they are known to leave the scene forthwith. And that includes many encounters island wide."

"Once you tune into the animals natural communications system, in other words, 'synchronizaTion' followed by a transmitted 'passkey,' then the situation becomes positive, never in our experience negative. Nevertheless, we do keep all food sealed, especially because there are numerous bears in the area and the researchers are often away from camp or asleep," I add.

Kirk switches subjects saying, "Does everyone see the lodge on the pond's far side? We call that a bearing of twelve o'clock from this meadow. Four adult beavers live there now."

"Why not transmit the arrival message and get Justin and Nick to run parallel computer programs?"

"We're both here and ready," announces Nick.

Kirk had previously placed a large glass soda bottle half under water in the mud and half in the air. He reaches out and at Justin's practiced commands he raps the bottle on the neck with a meter long tent pole. The duration between raps, or the 'alpha concept,' is 32 seconds. After two minutes Kirk puts in a syncopated signal one second early, followed by an identical 'onTime' rap. There's not a sound or a stir on the glassy calm pond until - - -

"Moose! Nine o'clock, near the other meadow, heading into the water for a drink, female!" as Kirk points and silence reigns, except of course for George's click, zip, click and the equally low air sounds of the bottle taps!

"Now switch to the addition of two beats with syncopated notes, miss the next two and then repeat that cycle without losing the 'alpha rhythm'," I suggest.

The moose, after drinking, looks up with mild curiosity. Almost simultaneously two beavers emerge from the lodge and swim toward us.

"That's Bonnie and Clyde," announces Kirk. "Clyde is the larger one, now at eleven o'clock."

Very quietly Justin counts down to the mark and Kirk shouts, "synchronizaTion on Clyde."

"SynchronizaTion on the moose at the same time Nick. You record her. Miss the next 'onTime' and signal on 'lateTime,' Kirk."

"As planned maestro. You watch, it is usually Bonnie that returns the passkey," replies Kirk.

The greeting goes out and we get an 'overlapping reciprocal greeting' from Bonnie as she uses quick head motions for signals. To our combined amazement, the moose also returns the 'greeting,' but not overlapped in time. Then, astonishingly, there is an effigy of effigies for George; Clyde climbs out of the pond beside the moose and with each making quantities of visual and audio signals they seem to be exchanging possibly vast amounts of confidential information. Such is the possible nature of 'Rhythm Based CommunicaTion' in its evolved, interspecies, finest form. Soon Clyde starts to chew on nearby alder bark and the moose saunters west into a spruce forest.

"Try asking Bonnie, 'How many young ones'?" I suggest.

"Roger," says Kirk, as if still on the radio!

This time Justin's computer conducts the music of the next four transmitted signals. Nicholas has a duplicate backup recording of all the results while instructions are given to the other students on how to prepare a 'beach feast.' Then every seven to eight seconds, for four times, Bonnie slaps her tail, not as the usual threat of danger but as a human trained 'counter,' to answer our question.

"Because those four slaps all fell within the two second windows, that is 'onTime, late, offTime and early,' we can have confidence in the correctness of Bonnie's answer. Probably each female has two young ones in their lodge."

"This is an amazing human penetration into the depths of Nature!" proclaims Hans.

There is satisfying silence as we notice ripples of rainbow trout which begin to stalk flies under the overhanging branches of the imposing and dignified, towering, fir trees near our meadow.

We wander happily to the barbecue where 'Village Inn delicacies,' as well as fresh capelin are sizzling on their shish kabob skewers and the brook water is as pure as the diamond like jewels that we are uncovering in our human animal communications research.

Turnstones Turning

Along our shore came a flock of turnstone[14] birds,
Stopping on a sandy beach nearby.
A leader, 'Alpha,' began to feed on slugs,
And many, many colleagues followed suit.

Then an eagle came but high above,
And Alpha quickly led the flock to flight.
They turned together all eyes on Alpha's wings;
They turned in seeming perfect synchrony.

Using high speed film we wondered how.
We filmed the wings and then took careful note,
That Alpha's wings went out of synchrony,
Before each turn - a 'eureka discovery.'

If Alpha's late on downswing, then turn left,
If Alpha's late on upswing, then turn right,
If Alpha is an x 'duraTion' late;
Turn y degrees, on very next beat!

The main discovery is that such animals,
Combine signals, as upswing and/or downswing,
With rhythms, as an x 'duraTion' late.
Their communication indeed involves 'RBC'!

The 'Time' of 'RBC' is now defined,
'Perception of Lateness Relative to OnTimeness,'
It's a novel temporal concept called 'RBT,'
Different from most human communications.

---

[14] Ruddy turnstones (Arenaria interpress)

CHAPTER 5

"Nature-Contact"

After this informative and festive 'Eagle Beach' extravaganza, we return to Trinity where a short cruise summary precedes our landing and, via radio contact, a welcoming party gathers at the Inn, alerted and anticipative. Chris has learned to judge an expedition's success by both verbal and visual expressions as participants make their front door entrance, dressed like space walkers in their orange, Mustang exposure suits.

"Unbelievable!" says Elliott, with his handsome, statesmanlike, leader's style. "The marine scenery is the best I've seen, including the whales, eagles, bears, beavers, moose, seals, seabirds and shoreline. And Trinity, 'The Jewel of Canada' is priceless, as seen from either water or land.

Following close behind is George who exclaims, "And I may have captured the best of it on film with breaching humpbacks and bald eagles beside hundred meter ridges! Awe inspiring would more literally describe this morning. At this rate, I'll need lots of additional film for the expedition ahead."

Our manager, Sandra, smiles from under a massive mahogany carving of a blue whale, hanging above the front reception counter, "No worries about that George, we've plenty. We estimated that Dr. Nokatora, from Tokyo, took ten thousand pictures in three days."

Then, modestly but imposingly, in comes Alex. Chris braces herself to learn how a great scientific intellect, this master of physics and master of biology, will react to the new realities of 'Rhythm Based CommunicaTion,' the unfolding treasures of Newfoundland and the emerging new knowledge of the world in which we all share.

"An Ode To Joy," says Alex. "This was Beethoven's insight into natural music, into the true and long overdue scientific union of humans with Nature! It was the practical reality of my fondest dreams and more, when we first conceived these ideas so many years ago."

Chris smiles and begins to earnestly feel increased confidence in her husband's absorbing ambitions.

The last to arrive is Hans. Staff and students are nearby as he enters the main doorway, and in a resounding way, he echoes, throughout the Inn and beyond, just the one word, "MAGIC" - - - "MAGIC!" Hans calmly but majestically wanders past everyone, "Magic" - - - "Magic," and back to the 'Mustang Room' he continues, "magic" - - - "magic!" Elliott, George, Alex, Nicholas, and students now join him in emptying pockets and shedding their comforting, floatation suits for a well deserved quiet contemplation time.

Timothy soon transfers all the data from the portable Psion computers to our desktop PCs which are actively analyzing and printing out the morning's results, while Sandra makes copies of the *Ceres* data for a debriefing gathering scheduled for mid afternoon. Dody is busy with the preparation for tonight of a deluxe 'jiggs' dinner, including her famous peas pudding, made in a stocking-like sack and cooked in the same pot as the salt meat and vegetables, including carrots and the greenest cabbage. Below and in one of the large ovens, is being cooked the finest Alberta prime rib, for those preferring a 'CFA' modification. Our four distinguished guests all carry this popular acronym, which stands for the marvelous local description of, not a 'foreigner,' but a "Come From Away." Perhaps, however, they may prefer the unique salt meat flavour!

The favorite of all hors d'oeuvres is recipe number 113 for fresh squid, also called "Kyushu Calamari." This preparation, marinating in white wine and molasses, was given its distinctive number in honour of our record humpback breaching sequence, and the recipe was chosen from an even greater number of authentic Japanese cooking methods! When asked about the seeming coincidence of the numerals we then have the perfect excuse to explain how it's our belief that humpback whales always apparently intend to breach in prime numbers!

After a high English tea break with Dody's special carrot cake, followed by a brief data and strategy gathering in 'Archive Lounge,' our main sitting room, we depart in two vans with Joanna as our research leader and Timothy pointing out important historical highlights of Trinity.

"Here's the sight of the first court of the admiralty held in North America and there, the first vaccination for smallpox," says Tim to both vehicles, by using our quiet FM portable radios.

A few minutes later, "Mobile one to mobile two, over."

"Loud and clear. Where are you guys?" says Tim.

"Turning off to 'Skerwink Trail,' we'll wait by the 'Rabbit Path' entrance."

"You could go on to 'Ave Maria Lookout,' and count whales. We'll catch you there as we're trying to break the three minute tire change record! Oops! Standby one. - - - We just lost the four wheel nuts in a swamp!"

"Roger, take one off each other wheel!" as I log the time in my field notebook.

We park in a serene, light-emerald green, grassy meadow which abounds with pathways of tiny field mice or voles, amongst the myriads of glistening green, partridgeberry plants, sprawling crowbery and creeping juniper ground covering.

"Regardez! Here is 'main street,' there, 'upper water street' and everywhere are vole holes!"

"What sort of rhythm would one need to start human-vole communication?" asks Edward, one of this summer's two students from Britain.

"Probably a signal per second but it may be a few years before someone can learn to record data at such a fast rate. You could watch for snowshoe hares; we do work with them with a more reasonable 20 second rhythm."

Oceanside hikes in this part of the island are a great joy of botanical diversification. Lichen centred ecosystems abound on human size boulders carried down during the last ice age. I explain 'symbiosis,' showing an alga and a fungus locked into a gripping geometric cooperation, with the former gathering energy from the sun while the latter draws nutrients from the host. I suggest that for the lichen there is no 'future time,' no creativity, no future planning. And thus by logic, this alga is giving to the fungus, and visa versa, without expectation of a future reward.

"That's the heart and soul of 'altruism,' your favourite English word," says Elliott.

"Not by my training," replies Hans. "It is just symbiosis, a simple interaction between two organisms living in close physical association."

"A good debate, but the crux is the missing scientific proof that there is no expectation of a specific future reward. But as whales don't demonstrate 'future time,' and our results are similar for eagles, beavers, bears, and other animals, then how could we envision that plants would be able to expect distinct future compensation? The key here is specific payment, as with say human valuables. And they seemingly do this in their 'NowTIME'."

"Well, what does our expert say on that one, Alex?" asks Elliott.

"An interesting thought, but you know something chaps, I'm not an 'ex-pert,' for in the Russian language an 'x' is an unknown quantity and in your jargon a spurt is a drip under pressure!"

"My dear sir, in our language I'm not a chap!" growls Joanna. And the laughter dies away!

"Oops, pardon me! Is that what you say to take it back?" replies Alex. "I humbly apologize. May I include everyone with the much better word, 'peers'?"

From the lookout we see not one, not two, but many fin whales moving together almost as a unit, northwest along the English Harbour shoreline to our east. The blows are straight and high, more like royal palm trees, than the rounder bush shapes of the humpback blows of this morning. Standing still, like human statues, we marvel at the animal's close proximity to one another, while they swim almost as if they are a single synchronous being.

"What's up" thunders Timothy, almost jolting some of us off our rocky pinnacle.

"Good, you're here. We've got 'fins' at a range of a couple of miles," as I point the bearing.

"Great! Another car flipped our hub cap including wheel bolts which literally 'flew' into a swamp! George is just behind, so we can soon head on when you're ready," replies Tim.

Down from the lookout, we move into what Joanna calls her favourite woods, a place she has named the 'Regal Rainforest.' We are on a trail most certainly used by the Beothuk peoples centuries before the European's arrival. We pass beneath a shear cliff with a hundred tiny water sources, each one the creator of an entire miniature ecosystem. There are jade and juniper green mosses and distinctive deep forest lichens. I briefly imagine a return in the next life as perhaps a tiny insect, a life in such an apparent ecological paradise! Would I not need a sophisticated communication system? Could there be one here, right now? Joanna leads on while I muse for the moment. She signals for us to move quietly, for we are approaching our truly fascinating human-fox research headland. The spruce trees sheltering the darkened forest fade to smaller foliage as we huddle like a rugby team, ready to enter upon a realm of red fox[15] communications. Joanna and Tim, each with programmed 'Palmtop' computers, crawl ahead to the main den entrance where they suddenly stop in the center of the spacious, majestic meadow. When a signal is given, the remaining team moves stealthily in single file to the highest southerly observation area, where a wall of bush-branch camouflage shields us from the main research players, now lying at the den site below. The ocean view from this prominence is truly breathtaking.

Timothy produces resounding, sudden sounds with two rounded beach stones that have been left on location. The rhythm, as conducted by his computer, is one sound every 20 seconds. Nothing stirs save hundreds of kittiwakes bustling about their nesting colony on a jagged rocky island just below. Beneath Joanna and Tim, a dozen, young foxes, nine red and three black, are aroused by the tapping 'rock music,' sounds with a ring of familiarity.

"Listen and watch for any signals synchronized to Tim's beats," I whisper just loud enough for only our compact higher group to hear.

Nothing happens. Silence reigns except for the 20 second stone sounds, the cries of seabirds below and an insignificant wind whistling in the tall grass. I use hand signals to Tim to suggest continuing this conditioned greeting. Joanna, with dedication, observes the three den openings before her.

"Listen again now for any signals. I would train your binoculars on the middle opening. - - There!"

Joanna's elbow rises, indicating success.

---

[15]   Red fox (Vulpes fulva)

"Possible synchronization, from inside the den," I exclaim quietly.

Hans and Alex focus back and forth from their wrist watches to the den sights while George adjusts his tripod. Elliott and Edward, look to sea, in the direction of Green Island.

"OffTime," I say as the first message signal occurs. "OffTime, again" for the second message signal. While glancing at my watch, "The last message signal should be right - - Now!"

Then out of the centre den emerges the strongest and largest red fox kit, a handsome animal named Chuck.

Following him comes Andy, a smaller black male, then Sarah with a black tip on her left ear and from another hole emerge William and Harry, the smallest, each no larger than a petite, pet 'gumby' cat. George is in heaven as he photographs William, then Harry climbing onto Joanna's back! Tim continues with coded information transfer between himself and Chuck, while no other fox makes meaningful synchronized signals.

"Where are the parents?" asks Elliott.

"They go off hunting at this hour and bring back the bounty at sunset."

"You've just got to have an Edward underground there somewhere!" comments Ed.

"Edward the fox, me son, never shows up on time and has a blond tip on his tail!"

"No way! You've got your names mixed up. That sounds like Peter fox!" replies Ed.

"Not so, Peter is a wolf! Watch the other den now. Yes, here comes the dominant female, Diana, followed by - could it be? - yes, the almost all black Edward, the cunning fox!"

Elliott, who has been occasionally glancing back at the ocean announces - "Blows off Trinity harbour!"

"Fin whales, they're moving into the 'Skerwink Restaurant,' just off that nearby headland," I reply.

Diana, a leading fox kit, playfully wrestles with her brother Edward while William and Harry, the seemingly most playful, are chasing each other's tails in rapid, spinning circles. Chuck, a present 'communicaTions' leader, concentrates on his information transfer with researchers Tim and Joanna, both operating parallel programs. Our New Zealand contingent, Mark and Jay, are assigned to accurately make a maximum count of the fin whales as well as to estimate their relative lengths. On elbows and knees we three quietly creep our way to the highest vantage point beside a hundred meter sheer drop to the ocean below.

"Start your count anytime and then restart it as soon as there's any faint possibility for one whale to surface so as to be counted a second time. Your written length measurements should be relative, not absolute estimates."

"Shall we record locations?" whispers Mark

Yes. The twelve o'clock bearing here is assigned to the top of Skerwink rock; the range of the whales is now about a thousand meters."

Behind us, a third black fox emerges. It is smaller than the other two and seemingly shy. Harry diverts for a moment to roll him over and a fragile shriek cuts through the calmness of the human-fox communications causing Chuck's attention to be diverted from Tim's message. Immediately Sarah takes over and responds in the affirmative to the computerized question, "Enough food?" Chuck goes cautiously right up to Joanna where they exchange a variety of signals. 'Jo' (short for Joanna) has learned that these precious research animals all seem to be curiously attracted by her moving fingers in the grass.

Meanwhile I glance back to the ocean. "They're up," I say while counting the fin whales, from one to seven, and then I start again. One to five, one to nine! "Bearing two, range eight hundred, heading off shore," I disclose, for all in our group.

Alex is now on the upper pinnacle and asks, "What's the third black fox's name?"

"No name yet," I reply. "How about we name him Elliott, he certainly has charisma?"

Alex nods approval and then murmurs, "What's your guess for whale behaviour now?"

"They're up again!" whispers Jay.

"Notice they're turning in a clockwise circle back toward a known capelin restaurant off that headland at three o'clock," as I point the bearing for Alex.

"Nine, I saw a record count that time," exclaims Mark. "There seem to be three blows much smaller than the rest. Would these be calves?"

"Just a moment, Mark. There's Nicholas on my radio, talking to Sandra."

Elliott has now moved up. Hans's attention is glued on the fox experiment as I signal to Jo to escort him to the den sight.

"Land base one to *Ceres.*"

"Nick here, we're in Trouty Bay with a 'slinkie minke,' over."

"Roger, nine fins off Trinity, looks like their feeding corral is about to start, over."

"A-OK! Coming! Over and out."

"Elliott, may we name the third, smaller fox after you."

"Why not name it after one of my children?" replies Elliott.

"But it's sly, like you!" adds Alex.

Hans is now down with all three younger kits and George is having a photographers 'feast' from behind the blind. Tim just keeps cranking in the data, first from Chuck, then Sarah and now from the all black fox, Andy. It is the same where all animals apparently listen to the rhythm base and encoded messages but only one 'communicaTions' leader, at a time, responds. In the distance, we hear the powerful engines of *Ceres* at four thousand revolutions per minute with a presumed speed on these calm waters of 80 kilometers per hour! I suggest that the remaining students come to the whale observation pinnacle.

"You may see one of the rarest sights in all of cetology. Nick is coming in to photoidentify the fin whales and document the leadership in their feeding sequences. They're now just off the nearest headland."

"What's cetology?" asks a newer student.

"The study of whales, dolphins and porpoises or cetaceans from the Latin cetus for whale. I've been a full time cetologist, amongst only a few hundred, worldwide."

"Then why are you studying foxes, not to mention eagles, moose, beavers and bears?" asks Jay.

"Notice that I said 'have been.' We're well into human-animal communications research now so you may say that we're diversifying! I'm still mainly a cetologist, and that I suppose is for life!"

We watch in wonder as all fin whales form a clockwise circle, with those near the surface showing their enormous flukes, vertically cutting the surface like the dorsal fins of large male orca whales[16].

On the far side of the corral, we occasionally see the long, white, lower right jaws which are forcing the capelin upward and inward to be spun around like a giant whirlpool. *Ceres* arrives!

"I suggest that you paddle in from a hundred meters Nick and photoidentify the largest first, over."

"Roger."

Capelin are leaping clear from the sea surface in the corral centre. I now recognize one whale, a male over 35 meters in length, and swimming in water of a depth less that 70 meters. We believe this to be perhaps the largest fin whale in the North Atlantic ocean and one that we've studied for many years. George has moved higher where he can focus on either whales or foxes and with half zoom he claims to have just caught seven fins, all at the surface simultaneously.

"We need the left side dorsal of the second largest, the whale now at your twelve o'clock Nick, over."

"Roger, they're all on their right sides still, over."

George has heard the radio talk and he zooms in as the very whale in question turns dorsal side up to breath, and on our side of the corral. Alex and I, both with binoculars, are now with fixed gazes on the left side of this large dorsal fin and there,

---

[16]   Orca whale, was killer whale (Orcinus orca)

behold, is a deep diagonal scratch, an unequivocal field mark for life! With my radio on I speak to Nicholas, Alex and others at the same time.

"It's Scratch, a whale you named about ten years ago Nick, over."

"Great! We'll try to identify the others, over."

"I should have two or three shots of that fin, Peter. The sun is backlighting but it shouldn't hurt the scar resolution," says George.

"First rate research, 'me son.' Try for any other distinctive field marks, please."

Then, with little warning, a very specialized form of whale feeding begins. Capelin again leap from the centre of the coral as all three smaller whales simultaneously reverse direction and swim counterclockwise inside and into the oncoming myriads of tiny fish. The adults close ranks by decreasing the size of the circle and quite amazingly increase speed to about 30 kilometers per hour, or a full human sprint! Fish fly and water 'boils' as Nicholas, high on the mast and with passion, explains actions to the team below, all of which we can hear from his radio. Whales then apparently switch places, with one and sometimes two larger ones on the inside, while capelin continue to leap skyward for their intended but futile freedom.

"Are these post spawning capelin'?" asks Alex.

"Probably not. This looks like true evolutionary biology, the survival of the fittest or in this case the non-survival of the unluckiest! For post spawning feeding on capelin, humpback and fin whales are known by us to exhibit far less energy."

Suddenly the curtain drops on this 'Shakespearean-type' scene and the waters become calm. Some fish abscond shoreward while all nine whales totally vanish from our view. Tim and Hans are back with full attention on the foxes, having looked up for the marine highlights. The evening sun is low in the western sky when suddenly a sign of moden man appears at the edge of the meadow. Quickly Mark dashes around the lower outskirts so as not to disrupt Tim and makes a lunging dive at Trigger, a local friendly dog, unfortunately miles from home. With his belt Mark leashes the animal and slowly, carefully brings him to the upper observation blind.

Hans observes the last of these actions and says, "Double leash, and don't let go fellows."

The fin whales surface together, slightly offshore, and Nicholas has estimated perfectly; he is surrounded! Those onboard must be ecstatic as they start the communications research. This is now a perfect after-meal, low stress situation in which the whales should soon switch from signals to rhythms. And with 30 meter blows Nick should have no problem staying with them, possibly until dark!

"Why do fin whales have a white, lower right jaw but a dark, lower left jaw?" questions Elliott.

"For a host of reasons which are mostly known, so far, only by the whales," I reply. "When turning clockwise as you just saw, they manoeuver more easily on their right side, which places the white side facing down, and as you also may have seen, they then flash the white to force the capelin upwards to the air-water interface. We expect to be able to interrogate these animals about this characteristic in the near future, by means of our communications research. Right now hopefully,

Nicholas is reinforcing our declarative, rhythmic message for white, lower jaw."

"Rhythmic communications should lead to knowledge about nature at substantial cost savings," murmurs Elliott.

"Distant eagle! Terminate, and den the foxes!" I shout to Joanna, Tim and Hans.

Our danger message of five short signals send all, but Edward the naive fox, scurrying back into their dens as everyone stands to stretch before the homeward hike. Suddenly Trigger breaks leash and dashes toward the last remaining but unsuspecting young kit!

"Hans, look up!" shouts Alex.

An outrageous, high-speed, naughty dog and a naive, young fox are about to meet as Hans throws himself as interference and then screams in pain while Edward, the fortunate fox, narrowly escapes the attack and vanishes down his den entrance. Aid rushes in; we learn of, and examine Hans's pain and bruises.

The stalwart Hans rises, then bravely declares, "Don't worry about me comrades, it's been an unbelievable day for unstressed Nature contact!"

CHAPTER 6

Ceta-Rescue

"**T**hree cheers for the whales!" exclaims George as we open bottles of dry white wine in the upstairs kitchen of The Village Inn, before the celebrated cooking of the calamari. Onions and green pepper are separately sauteed, the wok is almost red hot!

"Three cheers for the capelin, eagles, beavers, bears, and foxes!" replies Hans as frosted glasses are filled, after Alex checks for the passing grade!

"Here's to Hans for saving the fox research!" says Joanna.

"Bravo Hans!" comes from Tim and others.

"How do you know that it is saved, doc?" asks Mark.

"One quick phone call to the dog's owner did the trick, three cheers for communications!"

Plates are spread about the table and a large central platter boasts colourful decorations from our greenhouse. Nicholas arrives with smiles from ear to ear and we all anxiously listen to get a brief taste of his adventure.

"It was Scratch's 'heard' all right and I recognized Tail Notch, Orion, and of course G.L., but not the three smaller whales. We were with them for almost two hours and, - well, I'll let the data speak for itself," as he disappears through a concealed door to the penthouse above,

"Squid[17], according to the experts, must be cooked over two hours or under one minute, nothing in between, as told by Japanese experts! We here at The Village Inn, have chosen recipe number 113 which is from 59 to 60 seconds, no longer! First, I drain the squid from the wine, molasses and secret ingredients and then place our master clock, which measures 'linear time,' but in a circular way, in this prominent place. Then we ask Tim to cover up the fire alarm as steam, momentarily, will fill the room! Could anyone pass around the wine again? Everyone ready?"

"Go for it!" comes from many!

Into the red hot wok go twenty sectioned squid and immediately out comes the evaporating wine, the aroma, the steam! Mark is counting down as I stir frantically -- 40,-39,-38, the room fills with 'a fog,' -- 30,-29,-28.

"Some 'tic' boy," Penny adds as she delivers more wine. The Chablis is dry but Penny is sweet!

In go the precooked onions and peppers -- 20,-19,-18, in go some spices, and finally, with almost an explosion of additional clouds, the marinade. The alarm sounds! Penny screams while grabbing George! Mark and now Nick count, as they have done so well before, 5-4-3-2- off fire comes the wok and in one swift motion recipe number 113 smoothly slides onto the serving platter, at the

---

[17] Squid (Ilex illecebrosis)

mark! The alarm stops thanks to Tim fanning it with the door and George with his jacket.

Penny remarks as she departs, "You men cooks are all alike, much ado about nothing!"

"Ladies and gentlemen, praise the Lord, and help yourselves."

Good fun, good cheer, good friends avail in the quietness of tantalizing taste sensations.

"Now please tell us how Nicholas associated the right lower jaw of a fin whale to a rhythmic signal, if that is indeed what he accomplished," states Alex.

"How about after dinner Alex when our hunger stress has subsided and one can better conjure up our projected 'thought experiments,' an expression that my idol Albert Einstein would use?"

"But we're having this marvelous calamari and we should be able to handle just one burning inquiry, doesn't everyone suppose?" asks Alex.

"It's ok by me Dad, I'll give it a try," says Nicholas, as the room calms, except for the subtle sounds of serious wine drinking. "The rhythmic message is 1-1-0-1-0 where one is 'lateTime' and zero is 'onTime.' The association during the signal and for two minutes afterwards is a white submerged vertical plate on the starboard or right side of *Ceres*. The message 1-1-0-2-0 corresponds to a similar black plate on the port or left side."

"Wow, that's too much! Unbelievable, especially after this great day, this wine and this sublime and prime recipe, number one-one-three!" adds Hans. "Why don't we find out something simple, like when and why you named a fox Diana!"

Joanna handles that one. "It's a beautiful name, for a beautiful female fox, and person. Sorry that I couldn't do it in poetry Hans!"

"Jigg's dinner anyone and everyone, served below in the Diana, oops, I mean Victoria Room!"

"Why was the word 'prime' used with this recipe?" asks: Mark.

Jay answers, followed by their seemingly novel New Zealand confirmation, "The number 113. Get it?" "No!" "Prime number. Got it?" "Yes." "Good!"

After dinner we sit back with Cognac and Kahlua while Hans tells a story of helicopter skiing in the Rocky Mountains, his favourite escapade environment.

"Once my oldest son was powder skiing through the tall trees at breakneck speed, with the rest of the family and friends trailing well behind. Then, there was a sudden scream, and no sign of the lad! We slowed to follow his tracks, stopping by the tallest tree trunk which disappeared down a hole etched out by months of mountain winds. There he was five meters below us and upside down, hanging by his skies and wedged into the edges of this enormous hole! All that we could do, with light ropes tied together, was to lower down a small thermos of well sweetened coffee! But with the help of the Lord we rescued him within the hour for a truly great family apres ski celebration!"

"If that's the older lad that you had here for 'whale contact' a decade ago, he must have been in his early teens when the accident happened. Please tell us about the rescue technique Hans."

"Yes, it was the same fearless son! We dug three meters by hand but we had to keep our skis on at first, as there was no base. We tied clothing around the trunk, as fortunately it wasn't too cold in the woods. Then, one of us scaled the makeshift tree ladder for the last two meters to pull the lad upright and eventually out of that death trap!"

Joanna whispers to Tim, "I'll bet the 'one of us' was Hans."

George, in his business planning mode, is on the verge of asking about tomorrow, when suddenly, Lloyd, the same prominent fisherman who owns the salmon net in Spaniard's Cove walks rapidly into the dining room saying, "Peter, I've just come from Lockston and there's an enormous whale, grounded, by the trouting trestle!"

"As large, or larger than your boat Lloyd?" asks Nicholas.

"Twice the size at least, although it's hard to estimate in the dark."

"Thanks Lloyd, we'll be there in a few minutes. I'll take four plus Nick on the Ceres with the two silver rescue kits from the main lab. The others can go by vehicles leaving as soon as possible. Dress warmly and bring all available flashlights."

The dining room clears in an instant. Penny who is smiling, as she and staff will be relieved to head home early, then announces, "Some house must be on fire, the way they all left so fast. Perhaps it's this here place!"

Everyone arrives at the scene almost simultaneously. Tim, who is manning a roadside radio, reports that the whale's length looks over 30 meters! Nick worries out loud that it could be G.L. or Giant Leader, the largest fin whale that he has just been studying and for which we have spectacular communications and behavioural data over the past twenty years. We anchor the Ceres just deeper than the giant tail while inflating our smaller rescue craft. Nick and I move in for identification leaving two radios with Hans and Alex. We approach on the right side, that being the fastest way to identify the species.

"This animal is larger than G.L. Nick!"

"Nick to Tim. Can you see the white on the right lower jaw from in front of the whale? over."

"Negative, over."

"Here, shine the light on the dorsal flank Nick."

"Dad, this is not G.L. or Scratch. It's a male and he's just too big! Breathing is every five to six minutes!"

"All stations, all stations. Unbelievably, we have a huge stranded blue whale[18], the largest species ever to have inhabited the Earth."

"Wow, and I've never even dreamt of seeing one; they're so rare!" exclaims Nicholas.

"We had a smaller blue whale stuck in the ice once but we got him out with the help of an icebreaker! That's why conservation organizations in California donated the mahogany carving above the front desk of the Inn."

"Rescue one to Tim. If you will relay to all that large group behind you, I'll give you some additional information before we start the rescue, over."

"Roger, over."

"Everyone gather and you can hear the radio directly," Tim shouts to the three hundred and more, amazed observers now on the road embankment.

One of the earliest to arrive, and on her way to The Village Inn, is eighty year old Edna. She has just driven from New Jersey for the third time in three years because of her deep, 'tormenting' affection

---

[18]   Blue whale, (Balaenoptera musculus, meaning streamlined and strong)

for the 'Great Whales.' Little can this 'super lady' believe her eyes, that she is suddenly encountering one of the mightiest of these 'awesome creatures!'

"This is a mature blue whale. It is one of the most highly endangered animals on Earth so that means we must try to save it although there is no immediate danger to the animal at the moment. The enemy could be the sun and heat of tomorrow so in that case we must keep the animal cool, usually using bucket brigades. This whale is more than thirty meters long and weighs over one hundred thousand kilograms or well over one hundred tons."

We move back to the *Ceres* where I extract a four meter by two centimeter polybraid, nylon harness, a small cork fishing float, a boat hook, a sleeping bag and some black electric tape.

"Tim to Rescue one. People want to know if they can help, over."

"We're ok. We'll attach a harness, take blood and loose skin samples, sketch some pigmentation patterns and the dorsal fin notches, then see if we can move this baby off the mud, over."

With the float taped to one end of the nylon line, I poke it beneath the narrowest girth forward of the truly enormous tail. On the third try it pops through and rises into Nick's waiting hands. I tape the sleeping bag to the remaining line and we slide it under this, so called. 'tail stock.' A very special slip knot configuration comes next. With a bowline and half hitch, in a large loop at the leading end, we then run the trailing end through this loop and tie a separate but similar knot which will attach to the one hundred meter nylon towing line aboard *Ceres*. Taping the cork to the second bowline we proceed to the other measurements while Alex intelligently floats the end of the towline, using a life jacket, back towards the giant flukes. We then take a moment to show Nicholas one of the enormous eyes, the size of the largest grapefruit, gazing upwards from just below the calm surface of this inner part of Trinity harbour.

"It's scanning fore and aft every few seconds and constantly watching me!" exclaims Nick.

"That's a sure sign of high biological stress," I reply, as I place a stethoscope near the heart. "The pulse is not unusually high, however, it's close to one beat per second!"

We move to Alex's tethered life jacket, remove it, thread the towline through the harness loop, remove the small float and return to *Ceres* with the towline end.

"One of our scientists calculated that a blue whale this size has about 500 horse power, *Ceres* has less than half that."

"What do we do if the whale tows us backwards?" asks Alex.

"We've tied a 'Double Slip Tow Line' Alex. First we release one end of the long line which slips through the harness, then the entire tail sling, plus sleeping bag padding, releases itself from around the animal."

"Sounds like you've done this before. Let's give it a try," comments Hans, with slightly nervous shimmering in his voice.

"*Ceres* to Tim, 'T' minus three minutes and counting, over."

"One end fastens to the stem towing bolt, port side, Nick. The other goes through the mast block in a locked position and back 180 degrees. Four of you will have to hold the strain until I slack the throttles all the way to neutral, for release. Alex will you stand on the 'console couch,' hold on tightly, and shine these two lights on the mast strain and the tail harness, please?"

One powerful engine, then the other, start softly so as not to startle the whale; all loose equipment is stowed. We move away at a snail's pace and anticipation rises everywhere, both on board and ashore.

"*Ceres* here, 'T' minus one, all seems ready."

The towline tightens.

"Line on surface straight back to the whale," apprehensively comments Alex.

"Ready on tension Nick. Holler if you're losing ground."

The mighty engines ease upwards from one thousand to two to three thousand revolutions per minute with now to nearly full horsepower, but *Ceres* remains stationary. Another hundred, then another and it feels like the transom might suddenly eject from the vessel!

"One slight increase and that's enough!" I say.

Slowly, ever so slowly the great whale inches off the mud but the only indication on board is a thunderous roar of approval from the almost five hundred observers now on the shore. I slack gently back on the throttles and line tension seems to reduce somewhat from its close proximity to breaking strain. From Tim's vantage, the great whale moves slowly out into complete darkness and to freedom in the deeper waters. But what motivated this giant to strand? Will there be a recovery?

"Large circle left. We'll head him out the harbour mouth before release."

Halfway through this gentle turn, the whale begins to exercise some of its five hundred horsepower and all four heave back against the new strain.

"We're going to lose him Dad, turn faster," hollers Nick

I ease her more to port while watching the mast and its block, bending under the enormous strain. *Ceres,* in forward gear with over two hundred horsepower of thrust, stops in the water and then begins to go backwards with higher and higher splashes over the stem, splashing all four linesmen.

"Ready for release. - - - Wait. - - - NOW! And the towline flies out through the block while *Ceres* slows, roars ahead, then slows again as the engines return to neutral. The valiant crew, save for Alex, pick themselves up from the lower deck, laughing, wet, but full of joy! Nick and I haul in the towline and off we head to hopefully follow the whale. Street lamps light some of the way in this inner harbour but there's no sign of our 30 meter blue whale. We move to mid harbour and turn the engines off to listen.

"Blow, one o'clock. about one hundred meters," says Nick.

We pull up alongside and grab the harness with our boat hook as it conveniently slips off the tail.

"That was a bit of luck. I didn't expect to get that back."

This whale, whom several had suggested that we call Alex, slips downward in the central part of the harbour and is gone from our sight in the darkening night. We are soon home, exhausted both mentally and physically but brimmed over and laughing with deep, inner contentment.

The next morning I rise at 4:30, wake Nicholas and we prepare toast and fresh coffee for the unsuspecting core team. Elliott and Alex rise quickly with just a bit more resistance from George and Hans. Fed and suited we're away by early light and search the inner harbour in case the blue whale was unable to navigate to safety. Sure enough he's in a western cove but luckily, gently grounded on a rising tide. We follow the same procedure with a little time for all to be intently 'watched by the whale' and at a half meter range from a scanning, giant eye. Elliott reaches gently out to stroke the animal, which we have named Alexi, to avoid confusion, and upon contact the whale undulates long rhythmic waves to both extremities of his body. Hans monitors the heart which has now slowed to a healthier forty beats per minute. Nicholas is preparing the tow lines aboard *Ceres.*

"Wow! What's that feeling?" exclaims Elliott while he crosses both arms over his chest with a look of total amazement.

"What's happening to us?" says George letting his camera drop on its neck strap.

"That, gentlemen, is the famous blue whale subaudio signaling known to some as 'The Twenty Hertz Monster!' Alexi is producing about twenty cycle per second sounds at the astounding intensity of about two hundred decibels, enough sound to travel to Bermuda and back!"

"But I can't sense that in my ears. It seems, on the other hand, to have totally filled my abdomen," comments Hans.

"Precisely, it is too low a pitch for your ear drums to handle. Alexi is vibrating your entire inner body. This's a phenomenal, once-in-a-lifetime feeling."

"There it is again! Sensational!" says Alex. "It's music to the soul!"

"Nick, can you feel the low frequency sound transmissions?"

"Sure do, maybe they're a distress call to another blue whale," shouts Nick from the *Ceres.*

We gently pull away from this seemingly magnetic union to continue with the rescue attempt. This time Nick skippers the *Ceres* while Alex takes up the command observation post afore the console. With rising tide the 'unstranding manoeuver' is simplified and on short tether we are able to aim Alexi straight out of the harbour mouth. Pulling slowly up along his right side after release we guide him away from Admirals Island and safely into the deeper waters beyond. While coiling the lengthy towline we see Alexi exhale to the southeast, perhaps on his way to a Grand Banks cheerful reunion. We arrive at the Inn just as does Dody, so a quiet order of a giant breakfast for six, with memorable conversation, is set in motion.

"This has got to be one of the highlights of my life," says Hans.

"And mine also," I reply.

CHAPTER 7

"Inner-Contact"

Stacks of pancakes with real maple syrup, brought by Elliott, and small Alberta steaks with thin slivers of kidney, large slices of bacon, eggs and Dody's special 'doubloonie-shaped' sausages, emerge to our hungry 'pack' of six, comfortably seated at a round table! When the conversation turns to Nick, who was describing Dody's special, Alex shyly speaks.

"What in St. Peter's name is a doubloonie?!"

"Two loonies," replies Nick as quick as a flash."

"And what, prey tell is a loonie?"

"Our new dollar coin with Queen Elizabeth II on the front and a loon behind," states Elliott concisely, while demonstrating such a coin for Alex.

"Did anyone notice the polar bear on our two dollar coin?" I also demonstrate. "Well, it's now called 'The Queen with a bear behind!' Oops, sorry Elliott," as a feeling of embarrassment permeates outward from his patriotic soul! Notwithstanding that he is of fine French Canadian stock, and Elliott has truly great regard for our royalty. "That was my strongest feeling of empathy; we must be coming closer together as companions. I do apologize from 'me art,' or rather in our Queen's good English, from my heart!"

"My strongest feeling ever, was what you called 'The Twenty Hertz Monster'," says George.

"I Agree. And the fact that we all had the feeling synchronously, as a unified and unifying experience, could well create 'bonding,' lasting perhaps forever and a day," I add. "If we have a reunion I'll wager a loonie that someone recalls that special moment. By the way, the description of the intense twenty cycle sounds of Alexi, the blue whale, is Navy talk, going back fifty years."

"This syrup is incredible; why, oh why, don't we have more maple trees in Newfoundland?" contributes Nick.

"Well you do have the maple leaf on your, I mean our, Canadian flag!" replies Elliott, which brings on a cheerful note, and signifies that our 'senior statesman' is back in the game!

Penny arrives.

"Yous boys want 'Moca Java, decaf' or just Java talk, like your brother Tim?" as she glances at Nick.

"Translation please?" confides Hans, close to my ear.

"Regular or decaffeinated coffee, anyone? Penny, why are you wearing the same wig as yesterday?" I gently tease!

"Least that's better than wearin yar old bald spot every day!" she counters.

"Penny, you look terrific this morning," adds George.

"Thank you sweetheart," as she mentally records additional orders. "And I even borrowed me son's new head stuff, especially for naturally curly locks - - like mine" with a nice but aggressive glare directly into my eyes! "In my culture ye gives what ye gets!" And then she's swiftly through a swinging door to a 'high tech coffee contraption.'

"In the humpback's culture ye gives and ye gives," I comment.

Moments later we sit back to savor the ambience of the moment, of the coffee, of the morning, of the fellowship, while Hans asks, "What is the supposed relationship between 'Rhythm Based CommunicaTion' and one's emotions?"

"According to previous research as well as the 'Lanzarote' paper itself, that is the published paper that I gave at a conference back in 1995, there now apparently may well be two distinctly different paradigms of communication, two classes that are capable of occurring simultaneously. You can additionally read of a fascinating fish and bird communication conjecture, in this same, paper. The two archetypes are: a, 'Signal Based Communication, SBC,' which utilizes the encoding of information in the sensory modalities, that is in: 1, sight, 2, sound-touch, 3, taste, 4, smell and 5, hot-cold, wherein this communication, energy is directed to one's central nervous system, that is, principally to the architecture of the conscious mind; and b, 'Rhythm Based CommunicaTion, RBC,' which utilizes the encoding of information into 'Rhythm Based Time,' or 'The Perception of Lateness Relative to OnTimeness,' also called a common 'synchronizaTion,' wherein such 'communicaTion' energy, may be directed to one's DNA and RNA, partly to a rhythm base that may well be, seemingly, in one's conscious and unconscious minds, and which is presumably highly related to one's emotions."

"Firstly let me explain that the use of an internal upper case 'T' in words such as 'onTimeness, duraTion, communicaTion, synchronizaTion,' and the amazingly important 'informaTion,' represents the inclusion of only 'Rhythm Based Time, T, RBT' or 'RhythmicTime, RT', and not 'Conventional time, t or Ct,' displacement/velocity or space/speed."

"Emotions are often feelings perceived in parts of the body other than the central nervous system and by the above definitions, in such cases, these emotions now seem principally the result of 'RBC' messages. The transduction mechanisms to convert from 'Rhythm Based InformaTion,' as the above type b definition, to action potentials, nerve based signals and the conscious mind, are largely unknown and thus great potential lies here for advanced research in biochemistry. So, assuming this to be an active aspect of comprehension, then emotions first received in the unconscious can be translated to readily understandable feelings associated with conscious thought. But this is not always the case. Sometimes one acts on one's feelings without knowing exactly why, which, as you know, we often refer to as intuition. This can happen in the 'blink' of an eye, as Malcolm Gladwell would say."

"On the other hand, messages received in the conscious mind are frequently linked with the unconscious, and thoughts seemingly stemming from the central nervous system are transduced into rhythmical 'informaTion' causing feelings, apparently coming from sundry parts of the body, and thereby correlated with one's unconscious mind. The Jungian concept of the 'Union of the Unconscious' thus refers to direct rhythmic 'communicaTion' between any two synchronized, unconscious minds, often not involving the transduction and thus not directly known to conscious thoughts of either organism."

"Thus 'Rhythm Based CommunicaTion' may become the principal discourse of the emotions, of joy, excitement, love and of many more. It is the peacefulness that transpires when 'A Great Whale' gently moves alongside, to watch, to explore and to communicate with the human species."

After a moment's reflection, I continue, "Let's meet in two hours so that we can study yesterday's data and plans for today."

However, with little notice, unusual wind gusts suddenly shake the northwest side of The Inn, including this 'Victoria' dining room, also now referred to as the symbolic 'Chamber of Altruism.'

Student Seminar #2 - Simple Biophysics

Students are assembling in the 'Eagle Room,' which fortunately is on the leeward side of the Inn and is thus far less effected by the rising winds. Chris approaches our breakfast table and inquires of everyone's euphoria, which she can sense in our eyes, our speech and our apparently enormous appetite as judged by the numerous serving dishes! I quietly depart, as George tactfully tells about this morning's amazing adventure of Alexi the blue whale, and how fortunate it was to have rescued one the most endangered animals on the planet. Chris can feel the electric sense of excitement and being of nursing training she is relieved that this indicated storm should give our group some quiet time to reflect.

As I enter the 'Eagle Room,' "Attention s'il vous plait! - - Merci. This second student meeting will try to explain new theoretical concepts about the past, the present and the future. I've already ordered the hot drinks so now's our chance for good communication, good questions and good application to your various animal experiments, and, quite possibly to both your remaining life, as well as your very inner being."

"Do you mean to infer that these discoveries in communications may improve our careers as well as our total well being?" asks young Jason.

"Precisely! Let me explain."

"Firstly, what of the boat trips?" inquires Ed.

"There is no known way for humans to carry out marine mammal, 'RBC,' research in the wild, at wind speeds this high so we'll concentrate now on other things, like pristine, unpretentious knowledge, Edward!"

"Animals, other than humans, do not possess what is generally thought of as semantic language. Apparently, an evolutionary selection pressure has not existed in order for them to develop either a language or detailed thoughts of future plans or future creativity. There are no significant cave paintings by non-human species! Animals have memory but their conscious thought 'seems' always in the present, and of the past. Migratory patterns, such as 'north to feed, south to breed,' say for our North Atlantic humpback whales, seem like generalized futuristic thought processes, but they are more likely caused by a genetic trigger based on internal biochemistry, external conditions, or environmentally based learning. But we now define the concept of 'future,' to consist of detailed plans and foresight such as are found, so far, only in humans. One's past is related to memory and in molecular form memory seems found in all living organisms."

A thought filled silence follows, then a short break filled with sounds of pouring and consuming hot chocolate and coffee.

"Now let's discuss: 'A New Logic of an Old Concept,' that of: 'TIME.' And in particular, I'll mention the mystery of biological aspects of the physical time variable, often represented in your past science courses simply by the lower case symbol t. Using a definitive definition of logic, the following opinions may approach principles governing correct or reliable inference, involving the human enigma of what seems to be an improved temporal theory."

"There now appear to be two temporal types: a, linear or cyclical 'Conventional time, t or Ct,' also designated as 'time' with the usual lower case t, and b, cyclical 'Rhythm Based Time, RBT, or RT,' in contrast designated as 'Time,' spelt with an upper case 'T,' or by the symbol 'T' alone. Additionally we suggest the word 'TIME,' with all upper case letters, be used to represent a sum of t+ 'T'."

"With this convenient notation we have the need for new words, with an internal upper case 'T,' such as: 'duraTion, synchronizaTion, communicaTion,' and seemingly the immensely important 'informaTion,' as well as and including closed words such as: 'onTime, lateTime, offTime, earlyTime, RhythmicTime' AND the equally important 'NowTime,' or 'NowTIME', where, as before, 'TIME' = t + 'T', all representing the incorporation of cyclic 'Rhythm Based Time, T or RBT,' or simply 'Rhythmic Time'.

Important differences between these two temporal types a, and b, are:"

"1a. A duration of 'Conventional time, t or Ct' is displacement divided by velocity where such displacements and velocities are external to a measuring, cyclical, working clock, or if the clock is linear (as for example a water clock), then such a linear clock must match, on a one-to-one basis, the said cyclical clock."

"1b. A 'duraTion' of 'Rhythm Based Time, T or RBT' is displacement divided by velocity where such displacements and velocities are internal to a measuring, cyclical, working clock, or are associated on a one-to-one basis with the internal rhythms of such a clock, as for example an age."

"To clarify these two novel statements, our Earth is a measuring, cyclical, working clock and if one remains in a stationary location then one's cyclical velocities are associated with 'duraTions' of 'T.' If however, one moves, then any travel time is in the normal durations of t. In communications science, if one synchronizes one's master, internal, biological clock, one's 'Suprachiasmatic Nucleus, SCN,' to any exterior 'Rhythm Based Time' cyclical, working clock then the resulting 'duraTions' become essentially 'internal' to such a cyclical, working clock and such becomes a potential starting point for 'Rhythm Based Communication, RBC'."

"2a. 'Conventional time, t' is unidirectional, counting only in the direction from the present to the future, as in radians from 0 to 2pi to 4pi, or in degrees, from 0 to 360 to 720."

"2b. 'Rhythm Based Time, RBT' or 'RhythmicTime T' may have a sign change at pi radians (180 degrees) and counts within its 'NowTime,' in radians from 0 to pi to 0 (or in degrees from 0 to 180 to 0). 'T' can be bidirectional, as when viewing the Earth from either pole."

"3a. 'Conventional time, t' is relative to space, depending on transmission and reception spacial characteristics (Dr. Albert Einstein)."

"3b. 'RhythmicTime, T' is relative to 'synchronizaTion,' depending on minds and mind locations, and can be Earth-Sun absolute, based on the Earth's rotations."

"<u>Both temporal forms are mental readings OF (not ON) various clocks."</u>

The inquisitive Mark then asks: "Do you mean if we drive south from Ottawa for 12 hours, that we also go east on the internal Earth clock for 12 hours, and end up where 'China Space' was?"

"Yes indeed. South is a 'Conventional time, t duration,' which can add to the east 'Rotational Time, T duraTion,' but it makes no sense to multiply the respective velocity vectors. There will be more on this important point, later."

"Peter, there's a phone call from England," interrupts Chris

"OK everyone. Review these ideas. They are also in these copies of Target Article 92 from the Karl Jaspers Forum, on <www.kjf.ca>."

Ascending the 'Escalator' on every other step, I enter our office piano living room and head for a comfortable high back chair at our master computer terminal.

"Ceta Research, Peter here."

"This is William. I received the e-mail with Alex's new calculations and you just have to verify the orthogonality probability of 'RBT.' I suggest that you teach Ida a pi/2 radian concept and immediately try for evidence of a significant response relationship of communication at pi and then at 3pi/2 radians."

"Ok! Superb thinking. But we are land bound now with gale force winds so I'll get Alex to reply 'asap' Wil."

"'RBT'does indeed satisfy some relativity theory as well as some quantum mechanical problems, provided it is instantaneously supersymmetric as you've suggested," comments William.

"Great!" I reply while confirming by keyboard onto an automatically addressed e-mail window.

"How are Chris and the boys?"

"A-OK as Alex would say!"

"What space have you if I come over?"

"I can give you a day's warning if you can be on standby. We've found what appears to be a large, stable, near shore humpback restaurant and all I need now is a reliable weather forecast. Confirmation and a math question are almost set for your 'inbox.' Thanks for the great idea."

"What's up Dad?" says Tim who rapidly snuck away to enjoy some waffles with the Ontario maple syrup, in our adjacent, apartment breakfast room. "Who was that on the phone?"

"One of the brightest minds on Earth and apparently working on Alex's discoveries."

"I can believe that, seeing as how you are still a bit shaky! I bet I know who it was!"

"Could you possibly prepare five handouts of yesterday's data and do a presentation in about an hour or so?"

"Now wait a minute! The data is almost ready for print and xerox but you'll have me shaking if I have to make a presentation to what we're all calling your four friendly super humans!"

"Well ok but come and help anyway please. And you may have to change that description to the 'mighty five' if William comes over from Heathrow! They're all rather human, y'know, just like Nick and Kirk. Remember Nick's cherished thought of Nelson Mandela, a truly great man, who believes that humility is the finest of all virtues. Can you see it in all of the four?"

I notice that Elliott has gone for a sleep and that George has taken a large camera to bits, which pieces are all over his giant bed, for a routine cleaning. Hans and Alex have started a chess game, a true 'survival of the fittest,' in the main floor sitting room, also fittingly called the 'Chamber of Evolution.' I can't resist stopping briefly to silently study the play. Thence I return to the student seminar.

"Gentlewomen, gentlemen! You've digested some of that 'intro' I presume, so lets talk about 'NowTIME'," I say while re-entering the 'Eagle Room.' Suddenly the main fire alarm initiates itself into full acoustic action! This incredibly loud (160 decibel) clanging is pure 'NowTIME;' it is producing piercing sounds just outside both our 'Eagle' seminar room and the 'Archive' chess lounge where Alex and Hans are in the heat of a spectacular, complicated battle!

"I'll write the 'NowTIME' ideas for home study. Take your notes, Nick take the flip chart, and we'll meet on Taverner's Path by the birch trees. Exits are front, back and the east side rear corner."

I race through the main dining area noticing massive quantities of dark smoke leaking outward from the kitchen. The emergency, foam jettison system is working at full force with horrifying, grinding, groaning surges! Racing up the escalator I meet George on his way down.

"Front exit old buddy, have you seen Elliott?"

"No. Where's the smoke source?"

"Basement or first floor. There's a phone at reception with the fire number on the right side. Could you call them please?"

Seconds later I'm trying to rouse Elliott when Tim enters his room.

"Is your mother awake?"

"Of course Dad, she's clearing rooms from the other end. I was told to go back for the Capo de Monte, the top desk drawer and the blue whale carving, if there's time."

"Elliott, s'leve, rise fast, we have a fire! Tim, you escort Elliott out the back fire escape and meet out front. The 'escalator' will be smoke filled by now! Be safe!"

Seconds later I try to enter the kitchen by the staff stairway but when opening the fire door just a crack, I am turned back by both pressure and thick black smoke. Returning upstairs, I meet other guests in the hallways of 'Chambre City,' head them towards the rear fire escape exit and inquire if Tim and Elliott have gone on ahead. Tim comes racing past me into our apartment where he's headed for his mother's paintings, other art pieces and our family photograph and document drawer. As I hold my breath I descend the front stairway at full speed to meet Nicholas detaching the meter long mahogany blue whale carving, now probably valued at more than the Inn. Six masked firemen enter the front, west door and I direct them through the 'Victoria' room toward the kitchen. Chris is taking a mental roll call on the front lawn as other firemen arrive and I lead them outside and eastward to the back basement entrance. There, I notice that Tim is stashing art work in the green house some thirty meters away and almost all of the one hundred residents of Trinity are gathering nearby, alerted by the two fire truck alarms. I return to the birch tree meeting area.

"Dody are all your team out safely?"

"I think so Peter but I haven't seen Donna, she could've been in the basement at the laundry."

Where oh where is Elliott I wonder.

"Chris have you accounted for everyone, including Elliott?"

"Yes, except for Donna. Elliott definitely came out with Tim. There he is, walking in the rock garden!"

"No water needed," hollers Wilson. Your foam carbon dioxide system took care of that problem."

"We're missing Donna, last seen headed with laundry to the basement, Wilson."

"Roger, we've got masks and air here so I'll get a backup and head down those steps leading from the bar."

The wind has suddenly lowered and I return to the garden where Tim is now coming down the fire escape carrying his own computer with an open box of photographs balanced on top! There is great news. Fireman Colin exits the basement with Donna and he brings other news that the fire and smoke are apparently confined to the main floor and sleeping quarters upstairs.

"Wilson just told me that the carbon dioxide system worked and that a kitchen fire is out," I report.

"A-OK," adds Alex as he has wandered back to try to help Tim with his stabilizing act.

The smoke clears and we return, albeit slowly, to some of our previous activities. Outside, it is now calm and cloudless with such warm weather that I say to the students, "Let's continue in the whale bone garden by the guest house."

"Continue with what subjects?" asks Hans.

"'NowTIME,' whale thinking, animal thoughts, everything except the future and the past," I reply!

Hans walks to Alex, probably for a chess discussion but just possibly regarding the student seminar. I return to the Inn where Chris is busy listening to Dody, Lory, Ellen and Penny. I overhear explanations such as,

"Oh, my dear, what a mess, what an awful mess!"

By now Wilson is investigating the fire source for his report, and I get the gist of a bacon fat, griddle ignition spreading to the deep fish fryer and thence across the floor literally chasing the cooks out the side door in semi shock but fortunately to rapid safety. The enormous stove, the walls and the floor are now covered 'knee deep' in char broiled foam!

"What ya been doon to us Penny?"

"Just synthesizing suds for lunch! And wit all the cookin effort you just gota eat it too!"

'NOW' is 'MIND,'
BIOphysics,
'MIND' is 'NOW'

A mystery of mysteries is the concept of 'NOW.'
But to comprehend 'NOW,' we must understand 'MIND.'
And 'MIND' is 'Essos,' (pronounced as 'Eee - sos').
It's one's 'Event Space,' a 'Sphere Or a Spheroid,'
A DYNAMIC volume surrounding one's thoughts,
But there is a 'non-Essos' volume within.
And such mental boundaries are created by critical,
'Mental Vector Processes, MVPs,' which are surely,
The 'Most Valuable Players' in the 'Game of Life!'
They are 'mass and/or energy' vectors attaching to scalar
labels,
And destined to one's 'Mental Thought Processes or MTPs.'

Now 'MVPs' travel via 'Conventional time, t,'
However they can be ANALYZED by 'MTPs' in reverse!
And behold, such a dual process defines mind's 'NOW.'
Which lasts in thought until one's 'Event Space' renews.
And because one's mental space can be so DYNAMIC,
Such reverse analysis can 'reach for the stars!'
Plus, thoughts can work with multiple 'MVPs,'
As pathways of ears, eyes and touch, can be synchronous.
And, 'MVPs' also come from one's 'non-Essos' volume within.
Such 'mass and/or energy' is generated by thought,
And back may come some memory, understanding and
knowledge.

Behold, so 'NOW' is the contents of one's 'Essos.'
And in such 'NOW' let temporal concepts be 'TIME.'
Such concepts are: 1) 'Conventional time, t,' and 2)
'RhythmicTime, T.'
For in one's 'NOW,' scalar labels of 1) and 2) can be added,
But NEVER, EVER, multiplied or divided!
Such now seems the BIOphysics BOTH of 'MIND,' and of
'NOW.'
So then, let us state some semantic review.
'Conventional times are t's,' and 'RhythmicTimes are T's,'
'Essos' equals 'MIND' and is created by 'MVPs,'
Which transport scalar labels to thought.
And t + 'T' is allowed, but t x 'T' is NOT.

CHAPTER 8

From Moose To Minke

Students and guests are gathered in the garden, surrounded by the bones of a giant sperm whale, by high blackberry bushes, fresh produce for some of the evening meals and by a myriad of lichens on various rock formations. They are discussing "the chess game of the century" played in 1956 between a 'formidable' master, Donald Byrne, and the finest child prodigy ever known to the game, Bobby Fischer. Nicholas sets up a board to show Bobby's famous eleventh move (Knight to Rook 5 !!!) and Alex proceeds to partially explain just why this is 'one of the most magnificent moves ever made on the chessboard.'

"There's a phone call from a fisherman in English Harbour," says Chris, from the Inn's lower back balcony.

"Ah-oh! That sounds like trouble," I reply while pulling out my cell phone and moving amongst some nearby tall trees.

"Does anyone see why the 'genius Bobby' creates a psychological advantage by looking like he made the craziest kind of blunder?" asks Alex. There is a thoughtful pause.

"By attacking the white queen?" inquires Joanna.

"No, not quite, that is common in aggressive chess," replies Elliott, as I return to the class.

"Do you know, 'Doc'?" Joanna looks inquisitively into my eyes.

"Yes, but that is memory, that is in my past and I will tell you so that we can 'zoom' into the present and then finish where the fire interrupted us earlier. Bobby moves his knight to his rook file making its power of attack diminished by up to 100%, and that is normally a blunder in matched strength, masters play. But now, as an exercise, can you find out how it is considered as pure power? Alex here, later played against Bobby Fischer and we both remember studying this stunning victory by a boy of thirteen, within a day of its conclusion in New York. Try to see how Nick's board setup, which we'll later leave in the 'Archive Lounge,' essentially won the game for the black side on this eleventh move!"

"Now let's 'zoom into the present' to tell you that unfortunately a three metre, juvenile minke whale perished last night by wrapping itself in a nearby salmon net. This could be a fine chance to see some anatomy, including ear bones, critical pigmentation patterns and more, so I asked the fisherman, a friend, to tow it over the few miles to our main wharf here in Trinity."

"I'll just say a few words about durations before we simplify our discoveries for everyone, including the younger folks, and then we can have a break for minke whale research."

"Durations are not vectors, because they do not have both magnitude and spacial directions, but they can be associated with vectors, such as in travel, which association has caused past confusions. Durations are but differences in temporal scalar labels. Human abstract mathematics created a 'theoretical' multiplication of vectors but all known, living organism minds, including humans, cannot meaningfully multiply or divide either vectors or especially any of their associated durations. Such multiplications caused problems in 20th century physics."

"As an example, drive a vehicle south from Toronto, Canada, for 12 hours. During this journey 'Lake Ontario Space' rotates eastward to approximately 'Lake Baikal Space,' in Russia. Now your journey of 12 hours duration south, plus 12 hours 'duraTion' east, lands you in a 'China Space!' But if you were to multiply these durations, the resultant 144 hours is meaningless."

"There is, in addition, a vocabulary which is easy and essential to understanding new temporal concepts, as all such concepts are always scalar quantities, like 'pricetags' on merchandise, and not vectors, like displacements or velocities. We can define four varieties of such scalar tags."

"Firstly we must review modern characteristics of part of organism mind, also called one's 'Event Space Sphere Or Spheroid,' acronym 'Essos,' pronounced 'Eee-sos.' The two 'Essos Edges,' inner and outer, are created by the production of 'Mental Vector Processes, MVPs' which are combinations of mass and/or energy vectors and scalar labels, destined to arrive at one's 'Mental Thought Process, MTP,' near 'Essos Centre' (please see Glossary)."

"Inside one's 'Essos' is subjective reality. Outside one's 'Essos,' and within an 'Essos Interior Volume,' are one's potential future and real past, the latter involving one's memory, knowledge, understanding, unconscious, culture, education and more."

"Let the scalar labels of 'Conventional time, t' and 'RhythmicTime, T' be 'timetags' and 'Timetags' respectively. The four varieties of scalar tags are as follows. The subjective 'timetags' and 'Timetags,' part of subjective reality, are within mind, within 'Essos.' Objective 'timetags' and 'Timetags,' part of one's objective reality, are ex-mind, 'ex-Essos,' either beyond ones outer 'Essos' boundaries as one's potential future, or, within an internal volume, as one's memory and more."

"Now for exceedingly important but relatively simple mind mathematics. Within 'Essos,' minds can add scalar labels. But, they cannot multiply them. Hence 't+T,' t+t, and 'T+T' within mind = 'Real TIME' where 'TIME' = 't + T,' but 't x T,' is meaningless, as described earlier in the Toronto to China metaphor. Similarly, colours and shapes, also being scalar labels, can be added, but not multiplied or divided."

"Please do not confuse the scalar characteristics of 'Conventional time, t' and 'RhythmicTime, T' with the 'transport carriers' named 'Mental Vector Processes, MVPs,' which can create 'Bioscientific Vectors' containing the scalar 'cargos' of t and 'T,' or their sum 't + T,' or other such 'cargos' as colour and shape. These 'MVPs' are simply mass and/or energy vectors transporting scalar 'cargos'. They seem the 'Most Valuable Players' in the 'Game of Life'."

"We let upper case 'TIME' exist only within 'Essos,' within 'Mind,' and it can involve t, 'T' or 't + T' but never a temporal product such as 't x T.' Such a product allows complications for current physics."

"Freeze everyone, FREEZE!" Through a front gate, and behind the guest house, not 20 metres away, saunters a monstrous mother moose, earlier named Mable, and trailed by a calf. "Everyone, make a very gentle finger snap, but only when I raise my right arm part way or every 32 seconds." Mable stops and seems nervous, probably with our unusual 'classroom size!' Two minutes pass with obvious anticipation and nervousness amongst the many attendees, and presumably also amongst the two moose.

Then I softly say the magic concept, but in this case the very true concept: "synchronizaTion!" The apparent stress lowers. The 'RBC' message of 'come forward,' or '2,0,2,0,' works and both animals follow the path east-to-west, splitting our group in about half, which direction seems to have been the moose's intended journey,

"Make a wider pathway to the Inn parking area, but slowly," I say.

"What is the name of the calf," asks Elliott.

"We'll need a name. We ponder. How about 'Mitch'?"

Now the two moose have numerous folks on both sides but additional 'RBC' seems to be working and their stress remains low. It is so low that in fact a few of the younger ones are able to gently stroke Mable and Mitch as they pass by.

"George, if you go around the back of the Inn and across from our front doors, then you may be able to get a prize photograph of these two passing the Inn's entrance."

Moments later, "Wow! Was that a great 'RBC' demonstration for all?"

There followed many styles and kinds of affirmative replies, even a few tears of joy. 'Nature's joy' is surely one of the finest kinds!

"Now we can and should talk a bit more about the important concept of 'Now TIME,' or its equivalent 'NowTIME'!"

Charles then states, "Please explain one of your favourite sayings: 'ALL (real) TIME IS NOW TIME.' Otherwise temporal concepts are just 'timetags' and 'Timetags'."

As everyone, including our manager's two small grandsons Mike and Dan, gather on and about the grass, I puzzle how to efficiently convey this grand and international conjecture to such a diversified audience. I try.

"No one is negating the continued use of the conventional, physical symbol and variable that we all know as lower case t, which normally stands, and stands alone, for our concept of time. But this concept still remains 'the enigma of enigmas,' the mystery of all philosophic mysteries. The following enlarged concept of 'TIME,' is merely an attempt to solve a part of this enormous enigma."

"Conventional temporal concepts have both a past and a future. Such can be derived, mainly as intervals between events, from measurements and predictions which are scientifically very sound. This means that such measurements can be reproduced by someone else, at some other place, at some other time and with remarkable accuracy. Therefore we must begin our definition with time, or its symbol, the variable lower case t, and build, adding recent suggestions of biophysics, namely that humans may be the only organisms on Earth to have evolved substantial future concepts. So our mystery must include a larger group of ideas for which a symbol, upper case 'TIME' representing 'NowTIME,' has a more comprehensive, real and reproducible meaning, with animals as well as with humans. To discover this symbol, we must include another, a second, an independent concept of time."

"For you Dan and the other school kids gathered with you, what I basically just said, is that we now believe that there is a new additional type of 'TIME.' The older, common and well known first type is on your watch, is in your head and has a past, a present and a future. The newer, additional and less well known type is in your heart, in your feelings and it is always in your present. Now I'll try to describe this newer type of time, based mainly on what humpback whales have demonstrated in experiments right here in Trinity Bay and elsewhere."

"Let's proceed with the as yet undefined, biophysical notion of 'now' and its associated, seemingly valid, new physical concept of 'NowTIME.' The experimental discovery of 'Rhythm Based Time, or RBT,' is usually designated with an upper case T, only in order to differentiate it from the more usual identification of 'Conventional time' with its lower case t. 'RBT,' is defined as 'the mental perception of lateness relative to an agreed, biophysical, cyclical concept of synchronization, including onTimeness, between two or more minds' (please see Glossary). Experiments with many species of animals have indicated that 'RBT' is biophysically different than

'Conventional time, t.' This new 'Time,' which is used in communications for the encoding of information, by humans and possibly all species of other animals, is always in the present! So how does one enlarge the model of 'Conventional time, t'?"

"Imagine a real 'mind sphere' within each of us, on and inside of which all events feel in our hearts and minds to be 'in the present.' Outside the sphere on one side, the side of incoming energy, events are in our future. On another side they are in our past. So we define events such as my utterances, your cough Mark, your fly swatting Mike, as being in our 'NowTIME,' but only until the energy of the event, for example the energy of a sound, leaves the sphere, until one's next heart beat, next breath, next cough or until your mind switches to a different set of thoughts. Suppose Mark's cough energy traveled due north to Alex. Then we could label that moving sound with 'Conventional time, t,' the type that you know. As Alex hears the sound, the cough is real, the 'Conventional time' labels are real and both are in Alex's 'NowTIME.' But Alex has his own 'RBT' biorhythms which by definition are cycling, just like the wheels of your bicycle Dan, but these cycles can orient themselves into an east-west direction, as opposed to the incoming north-south direction! The new 'Time T' can be at right angles to the old 'time t.' But most importantly, in new communications research they work together like best friends! They can add to create a new, upper case 'TIME'."

"All events are associated with energy and/or mass vectors and those occurring outside one's 'Event Space Sphere Or Spheroid' (one's 'Essos') are by definition not in one's 'NowTIME.' Also associated with such vectors are 'timetags,' (or 't-labels'), mental, inscribed, machine made, etc., which are in fact the very numbers, and perfectly valid numbers, that we have been using to measure time, in all fields; 'timetags' (or 't-labels') can label vectors with 'Conventional time concepts'."

"From one's future such labeled vectors enter one's 'Essos' and it is then that the said mass and/or energy which penetrates one's present awareness, and their 'timetags' or 't-labels' become additionally associated with real 'NowTIME.' Upon leaving, one's 'NowTIME' the 'timetags' become part of one's past. The transition occurs when the 'Mental Thought Process, MTP,' at or near the center of one's 'Essos,' switches internal mental processing from one set of associated 'NowTIME' vectors representing a distinct event, to another set. An example could be 'conscious thought change'."

We then move into the Eagle Seminar room to avoid evidence of a temporary light shower.

"'TIME' is in the present mind of an organism now, and for any size of its 'Essos,' it consists of real 'NowTIME' concepts, real time t and its 'timetags' and real 'RhythmicTime, T' and its 'Timetags.' Outside of one's 'Essos' (both past and future) are valid labels, valid numbers, valid memories, valid plans, but they have not the 'reality of nowness' or of 'NowTIME!' Thus the submission that:

'ALL (real) TIME IS NOW TIME.'
Otherwise temporal concepts are just 'timetags'
and 'Timetags'."

Nicholas arrives riding my recreational racing bike and announcing that the minke whale is being tied to the main wharf.

"Wow! that was fast. The minke whale is at the wharf but we will tow it around to the big amphitheater beach about 200 meters past the theatre. Could Nick and Kirk prepare the *Ceres* with extra lines and a number two anchor, and I'll get the necessary vials and tools?"

About 20 minutes later we land the whale amidst many onlookers, our four expedition members and science advisors, residents and many coincidental Trinity visitors.

"This whale was named acutorostrata, or 'sharp nose,' along time ago and it may well be the fastest, when full grown. The upper jaw has the narrowest angle of all baleen whales, which here is about one hour on a clock or 30 degrees."

"Notice that the diagnostic flipper band has a distinct anterior white line but the posterior edge is fingered and milder in its contrasting colours. Also note the baleen plates, descending from the upper jaw, which now can be mounted with the skeleton as a museum specimen."

We find a capelin wedged in the anterior baleen plates which is not unexpected and we can see the extension of the lower jaw like an immense piano accordion.

"Can we get one of the baleen plates for a Newfoundland treasure," asks George privately.

"We could soon visit an abandoned whaling station where there are lots of dried baleen plates."

Nicholas takes a tiny skin sample for DNA analysis and then, after photographs, we prepare to sink the animal, temporarily, in about ten metres of water, with a heavy anchor. Students will then be instructed to prepare the skeletal parts, wrap such in a one inch mesh net and file this valuable package under a friend's dock on the opposite side of the harbour. In the fall we will then insert starfish, sea urchins and other welcome predators to process and clean the skeleton for a museum request from central Canada. The remainder of this seminar, including Alex's concluding remarks 'can be found in' Appendix III. Comments are most welcome.

CHAPTER 9

"On The Move"

Kirk arrives! Justin leaps from the van and straight away announces, louder than I would have desired, that after the wind storm Ida and Andrew headed toward Bonavista! In a quiet conversation Kirk explains the directional, 'RBC' messages from Ida indicating the two received concepts of 'departure' and 'eastward,' and that their boat had then followed the whole 'heard,' leaving Trinity along the towering, northeastern cliffs of 'Whale Alley.' Apparently the gale force, offshore winds had driven their food fish, the capelin, into deeper water, and it was now the humpback's time to search for other food further to the north. This early exodus caught us by surprise, having planned long before for a greater 'home field advantage!' But remembering my talk with William, I guess that moving was 'in my bones,' an unexpected, natural, 'NowTIME' event, pulling our human team together, strongly and dramatically into the present.

"All experiments on land mammals and eagles proceed as planned except, would Nick take Kirk and the four students, that are new to 'whale contact,' as well as food, fuel, and radios, and then move the *Ceres* so as to spend maximum time with Ida, as well as to migrate with her 'heard?' We'll try to be on the water for a 6 p.m. to sunset expedition, at your nearest port for expected, undemanding contact."

"What about the fire damaged packed lunches?" says Jay, who is scheduled to study red fox kits with Joanna.

"Make a salad from edible forest plants, a fire for hot 'Labrador Tea,' and protein from the emergency boat staples!"

"He's got to be kidding!" mumbles his New Zealand mate Mark.

Joanna interrupts! "Yes he's joking Jay, we'll buy lunch at the store on the way out and probably have dinner at my place tonight. Work hard and I'll make your favourite - lasagne!"

"A-OK as Dr. Alex would say!"

"The fire trucks have both gone so you might as well finish your chess game Hans until we find out when Chris will open our dining rooms. There may be some smoked mackerel for lunch!"

Moments later I'm chatting with Elliott in the lounge, while Alex's board battle continues, viewed occasionally out of the corners of our eyes. Nick has set up another board with 'the game of the (last) century' and the begging question of what should white's twelfth move be.

"William called earlier on his cell phone from somewhere near London," I say to Alex.

"Did he sound convinced of a new 'orthogonal, bidirectional Time'?" alertly questions Alex. "I sent him my simplified version of his 'many equations hypothesis of reality.' I suggested ten equations for our ten variables: seven of time and three of space, plus an equation of state."

"He sounded open-minded. You must have drawn a convincing position as he wants to come over and join the expedition!" I reply.

"That's amazing. Will usually works with a penetrating pencil not a wild whale!" replies Alex.

"Well yes, but he did have a powerful idea that we should teach Ida a 90 degree or 'pi/2 radian concept' and then try to biostatistically relate 'Conventional time,' using 'non-RBC' signals, to zero or 180 degrees and 'orthogonal time' using just 'Rhythm Based CommunicaTion.' to 'pi/2 radian concepts'!"

"By 'non-RBC' signals I presume that you mean transmissions outside the four time windows that you have been using with humpback whales for some time now," replies Alex.

"Exactly. Outside the 'onTime, lateTime' et cetera, four second windows that we've previously demonstrated."

"First rate! Let's try that tonight! Wow!" comments Alex.

Check mate in five!" announces Hans.

Elliott and I move closer as Alex stares at his defensive chess position with intense concentration.

"He's had it!" murmurs a student as we move away so as not to disturb the master in deep thought.

"No. If I sacrifice my Queen Knight Pawn and advance my King Bishop Pawn I can avoid your well set trap," remarks Alex to Hans.

The game continues with five rather rapid moves, resulting in an even score and a mildly disappointed Hans. An upset here would be the game of his year and he did have the advantage that it seemed Alex had many powerful thoughts circulating throughout his amazing mind. I find it necessary to interrupt and suggest a short chess break.

"Regarding the telephone call, I told William that we would reply. My main computer e-mail, by the upstairs piano, is set up if you want to send a message to his office. You and William can invent some experimental methods, if possible!"

"Us two? That's your human-whale chess match! Our game is to do theoretical physics but also to be sure to understand the biophysics. I expect that the latter is why Will wants to come over," says Alex.

"How on earth did you and some brilliant younger scientists deduce that nature might encode most information into 'Time' instead of into signals, like humans?"

"We just read our data. Can you imagine wolves and eagles surviving on a few hundred communication concepts encoded in signals, rather than many thousands encoded in 'Time'?" replies Alex.

Alex sends off an e-mail in short order and returns to finish the chess challenge.

"Take that!" sternly announces Alex as he makes a powerful pawn move.

Looking at the battle it is hard to imagine the fortuity, sagacity or perhaps a Carl Jung 'union of unconscious.' Alex is back, not only surviving but with a good potential for a winning position!

A few hours of quiet time slip away consumed with leisure, fire problems, equipment preparation and the home cooking of a giant, special, wild game stew at a nearby house. The chess match, however, reaches an even end game and stalls, looking more and more like a potential draw, as we five enjoy an early dinner in the lounge along with help yourself, hearty burgundy, Dody's homemade bread and a piano intermission of someone's favourite song, Andrew Lloyd Webber's, 'Memories,' from the musical Cats.

Soon we are moving swiftly along the scenic highway east of Trinity and debating the glories of Mozart's 'Jupiter' symphony while a multi speaker compact disk player fills the Explorer with the grandeur of the molto allegro, fourth movement.

"This is when the young genius reached his prime," comments Hans.

"Agree precisely!" adds George.

"Look here! We also have the 40th symphony, written just before your 'prime' Hans! Let's compare the molto allegro of its first movement, a Russian favourite," interjects Alex.

"Please excuse the technical interruption folks," as I switch on a portable radio. "Ocean Contact, Mobile One, to Ceres."

"Ceres back. We're off South Gull Island, over."

"Roger. Elliston then at 1730 hours. We've got your dinner, over."

"Great! Nick out."

I place the radio on the dash, smile towards Alex while ejecting the 'Jupiter' disk.

"You be the judge then Elliott. Here comes another masterpiece."

Now we are traveling through 'Moose Alley.' which is said to hold a Canadian record of thirty one sightings along a single mile of roadway! The claim solicits lookouts and an explanation.

"The large forested area on the high ground to our left is a moose 'restaurant' and playground, probably sustaining many animals in cooler shaded areas during our warmer summer days. Notice how there are wetlands, both ponds and a river, on our right causing animals to cross the road for a drink, usually in early morning or late evening."

As we slowly leave this hazard zone, suddenly a giant iceberg appears glistening in the blazing sun, just beyond the pristine islands off Melrose.

"The bright blue fissure is re-frozen meltwater from a calving crevice. It's just a few years old. The striking, white ice, right next to it, fell as snow in central Greenland well before the birth of Christ."

"Then what causes the turquoise colour beneath?" asks photographer George.

"That's our frequent water colour caused by a plankton bloom."

Stopping, we get out to admire this spectacular statue and wander near a roadside lookout while George saves the precious image for all to recall in the months and years to come. Suddenly there are faint low rumblings apparently coming from the direction of the 'Buckingham Palace' sized berg. There's a loud gunshot like crack, then another. The pinnacle starts to sway. Left then right, then left again. "She's about to roll!" I say. Ten more tense seconds pass as George switches cameras to a 'six frame per second' motor drive. The pinnacle moves to the right, then left again, each time with greater and greater incline. Next comes louder stress cracking, then the seemingly slow roll, followed by such a water splash that you could almost imagine a small meteor hitting the ocean! We stand in respect, in awe, in joy, before quietly getting back into our mobile world of musical masterpieces.

The penetrating first theme of Mozart's 40th symphony causes five speechless, happy individuals to contemplate diverse wonders of the natural world when Elliott breaks the language silence softly, "Great music! I'll vote with Alex; that way we're all tied up. Mozart's finest!"

"Wait a moment! What's your judgement vote?" states Alex having recalled my earlier smile.

"There's no contest!" I explain. "Whichever way I vote, when we get to Elliston, Nicholas will take the opposite view, so we'll still have the even split!"

"I'll take Tchaikovsky's, piano concerto number one, and I've heard our skipper playing a good bit of it," says Alex.

"Bravo! My favourite, and my Father's, likewise. The composer's name was also Peter, and it was thus, that I received my first name!"

We drive northeastward now at a good speed, slowing only for a snowy owl[19], pure white, flying then hovering momentarily over what must have looked like a possible evening meal, scurrying through a grassy meadow below. Down slides the rear window; into action goes a camera with an enormous lens, and the owl glides untroubled into the forest.

As we mount the hill entering Elliston, there is *Ceres*, both engines churning white water astern so that she literally 'flies' along the surface of the calm harbour, toward the main wharf. As we arrive at the dock, almost without thought I use a

---

[19]   Snowy owl (Nyctea scandiaca)

traditional, local, trap-skiff expression, "'Load and Go' everyone!"

"That's what they say here with a full net of fish, Alex. All hands load a trap skiff and go for the dock, then load it again and if it's a good day, again and again! You might hear a conversation where someone says: 'What fish me son?' This could be followed by: 'Load and Go' all day. And there's a fresh fish for everyone on the dock to take home for dinner."

"Is that true that all codfish are called just fish?" asks Hans.

"Exactly. If you see a boat coming in, and ask: 'What, did you catch 'me son?' You could certainly get an answer such as: 'Three salmon. two halibut, the rest is fish'!"

"Hey Dad! Where's my dinner I'm starved?"asks Nick.

"In the 4 by 4. Let's walk to the Explorer and you can fill me in on the biology." Then, looking to the unloading crew on *Ceres*, "Kirk, I have a Bonavista taxi coming to drive you and the students home. Dody will get you all some special dinners to celebrate the finding of Ida!"

Youngsters by the dozen, gather around the *Ceres* and George is busy arranging them along with their intrinsic thrills of being photographic models, being obligingly helpful and hoping deep within, to be on a magazine cover, a giant poster or to seek their worldly fame! George is amazingly equipped with a pocket full of prizes for every last observer! Everyone's mood is exceedingly positive as we 'Load and Go.'

Nicholas is last to climb aboard as Kirk, on shore, casts the lines away and we head out between partly grassy, inner islands, bleached in places with flat, surface nests and eggs. Circling above the islands are many glistening white, Arctic and common terns[20]. Next come tens of thousands of puffins[21], some on the water's surface in great blackish clusters, most, however, roosting at burrow edges, while others are well out of sight at ocean depths, gathering capelin for their growing chicks which are relatively safe on the towering South Gull Island. Beyond are the humpbacks!

"Main transmitter on signals every sixty seconds Nick."

Puffins burst to the surface on all sides as *Ceres* slows evenly to 'dead slow,' to give these amazing diving alcids plenty of opportunity to plummet again if they are in our forward going pathway. With the wind calming quickly some have great difficulty becoming airborne with four or five fish in their beaks, an extraordinary payload!

"Feeding humpbacks must have trouble avoiding so many puffins," remarks Hans.

"They are never found ingested by the whales, probably because puffins seem to fly better underwater than in the air! However, an unfortunate gull was once discovered wedged between baleen plates of a humpback in Norway!"

"When we get closer, watch for a whale firing a puffin into the air as they are too large to fit down the whale's throat!" jestingly adds Nick.

"OK team, we have nine humpbacks and at least three minkes or piked whales. Keep a good lookout for signals at the same time as Nick's 'count downs.'

---

[20] Common terns (Sterna paradisaea hirundo)
[21] Puffins (Fratercula arctica)

One humpback should finish feeding before too long."

"5-4-3-2-1-mark," says Nick as a misdirected puffin flies between Alex and George, standing less than a meter apart in the bow of *Ceres*!

Nick has asked Alex to scan between 9 o'clock (port beam) and 12 o'clock (dead ahead), and George between 12 and 3. Both men are wearing brimmed hats and Polaroid glasses to reduce the surface reflected light and increase contrast.

"Remember to keep the eyes relaxed and scan 90 degrees, or pi/2 radians in about 5-8 seconds in both directions," I remind everyone. "Think of your clock bearings so that you can communicate to the group, first bearing, then range, if a whale appears."

"Blow, 4 o'clock, 100 meters, heading toward us, it's Andrew," says Nicholas. "10 seconds to alpha. Mark! Synchronization!"

"Use the 'offTime' passkey, Nick." Then a kick tap from the mast signifies 'roger.'

"Please explain the communications," exclaims Hans from his starboard stern observation post.

"For Andrew we will use the 'passkey message' of 'offTime' twice, then 'onTime' once, to set the 'RBC' variables at 30 and 60 seconds for future messages," I explain.

"Seven o'clock VERY CLOSE," shouts Elliott!

"Heading our way, that's Ida, then Hubert, then Cecil, they've identified us for sure," I say.

"Ready on the first message signal 2-1-mark," announces Nick.

"Minke, 1 o'clock, crossing under the bow, watch for the white flipper 'armbands'," I say to Alex and George. "Three o'clock - one kilometer, gannets plunge-diving, watch for whales or dolphins driving the capelin toward the surface," as I point to our starboard beam.

"Second 'offTime' signal 2-1-mark, in sync. with the coded computer beeps," says Nick.

"The greeting will be complete, gentlemen in about twenty seconds so watch for signals in your segments, I'll take the forward half, Nick the stern."

The third and final underwater, acoustic pulse is softer, gentler, but right 'onTime.' Then miracle of miracles, Ida and Hubert surface together and exactly in the middle of our 'offTime' window with precisely synchronized blows, one on either side of Ceres. Then they signal together in the next 'offTime' window and Andrew joins with all three, exhaling together in the 'onTime' window. Such is a very complex but successful 'reciprocal greeting.' All the while we are proceeding east at dead slow using one engine; both engines are raised to shallowest depths. We now transmit the south concept, 'late, offTime, late,' and then turn the *Ceres* abruptly south. All three humpbacks follow. We transmit the north concept, 'late, onTime, late,' and then turn north with the whales following. We repeat these concepts again and then again with the same results.

"Now you must ask them if our east west coordinates are their 'Conventional time'," says Alex.

"That means abandoning the 'RBT' windows. What do you suggest?"

"I calculate that if you head east, you could transmit three pulses every six seconds at 6, 12 and 18 seconds, then west, using pulses every twelve seconds at 12, 24 and 36 seconds," replies Alex.

"If you are ready Nick, we'll turn east and west for three messages in each direction."

And the experiment progresses with detailed computer records but slightly less detailed human conscious understanding, and no cetacean mimicry.

"Now comes the crucial test Alex. We must switch to the interrogative and look for meaningful replies."

So we reverse the 'Rhythm Based Time, RBT' which is similar to flipping your watch over, and transmit 'early, offTime, early,' while going south. Ida answers in the affirmative with two flipper slaps in the 'onTime' window. We repeat and then we head north and transmit 'early, onTime, early' and again Ida answers yes! We repeat again and then test reliability by heading south and using the identical 'north message' and Ida, catching on quickly, answers no with two tail slaps in our 'offTime' window, meaning negative.

Then comes the big test! We head east with reversed 'RBT' and three signals outside our 'RBT' window variables to represent 'Conventional time.' Ida answers yes. Then we head west with signals at 48, 36 and 24 seconds ('RBT' reversed from the previous west message) and again Ida answers yes, this time with a half breach with two distinct flipper slaps on landing, just a few boat lengths away! There seems no forward going 'Conventional time' but instead variable, bidirectional windows of 'Rhythmic Time.'

"Looking good gentlemen; we must repeat the data and perhaps we'll get a different communicating whale, if Ida's stress increases due to hunger."

We try it again in differing order and each time Alex follows closely with efficient human-human communications aboard *Ceres*. Sure enough, Hubert switches places with Ida, then Andrew takes over the communications leadership. Each repeats Ida's replies. Anyone watching from the land must think that we are doing a geophysical grid survey for sunken treasure but I would have enjoyed explaining to them that we appear to be uncovering perhaps the greatest treasure of all, Nature's knowledge.

"If only Mr. Einstein could be here now I feel sure he would rejoice in this evidence for the missing variable of Nature, that's 'Rhythm Based Time' and its orthogonal, bidirectional and absolute properties."

"If only William could be here now, he's a modern day Einstein you know," says Alex.

We repeat the experiment on into encroaching darkness and then head for the safe harbour of Bonavista. Passing the peninsula headland I explain that we are closer to Europe, via the great circle route, than any land in North America and that this was not only the likely landfall of John Cabot in 1497 but also of vast numbers of trading ships during the following 400 years. It is also the location of the first satellite tagged humpback whale named 'Theophilus Argus' after both the fisherman and the satellite system used for tracking! We were able to sit in the Ceta-Research laboratory back at Trinity and track the whale hundreds of miles away. This was for both daytime behaviour where we have some knowledge over the past hundreds of years, as well as for night-time behaviour, where we have practically none!

In dead calm waters under the night lights of one of the largest fishing towns in the world, we speed toward the harbour, hearing whales in the distance and avoiding rocky shoals that are well known to Nicholas and myself. We entered Bonavista harbour just as had the replica of Cabot's 'Matthew,' in 1997, celebrating the 500th anniversary. However they had encountered stormy seas with a cold winter-like wind, but were welcomed by Canada's Queen, who had helicoptered in for the festive occasion.

Leaving Nick off on the first wharf he is able to get a ride to Elliston and deliver our 'Music Box' Explorer, almost by the time we've offloaded the precious data and extra supplies.

While waiting, 'the four philosophers' celebrate with the popular Newfoundland 'Screech,' a smooth, dark, mellow rum. Nick soon arrives and is assigned a most comfortable bed made of all six Mustang exposure suits with added chocolate and other energy foods.

"When the CD counter gets to 1320 we have the most fantastic embellishment of Tschaikowsky's first piano concerto, with stunning key shifts and rhythm," I say. "This creation helps to make him one of my very favourite humans ever to have lived on our planet, along with my wife, Newton, Darwin and Einstein, of course."

Nick breaks in with. "Of course? How about Nelson Mandela for his humility?"

"Beethoven for his sixth symphony," adds George.

"Did you know it is thought that perhaps 'Mr. Ludwig von' accidently discovered 'Rhythm Based Communication,' perhaps in his early teens, and if ever anyone deserves a prize for an important contribution to human knowledge, I would accordingly vote for him!"

"Verify that one please!" immediately comments Hans.

"A-OK! Opus 27, No. 2, or the Moonlight Sonata, has five, 'RBC,' greeting messages where melody notes come in an 'early, early, onTime' sequence, during the first movement, for example in measures five through seven. I'll demonstrate; we have the recording here."

"So that's a novel sort of one way communication I suppose," expresses Elliott.

"Exactly but modem musicians are now starting to analyze videotapes of their audiences to enhance the emotionally powerful and similar, but two way, 'RBC communicaTions' of music!"

"The department of cognitive neuroscience at M.I.T., in a recent book, uses the maxim that music is a universal language, but then the book goes on to say that as far as biological causes and effects are concerned, music is less useful, speaking in evolutionary terms that is," adds Alex.

"But we have shown now, and especially today, that rhythm is the starting point of what is probably an age old form of communication which takes organisms into 'RBC,' into another, distinct aspect of Nature and that the next steps, after 'synchronizaTion,' are rhythmic message mimicry or passkeys and lower stress which leads to greater survival potential."

"Don't forget that additionally 'RBC' not only produces, but also requires lower stress, as we've proven when hungry whales return to 'Signal Based Communication or SBC,' just before feeding," adds Nick.

"Correct, and we believe both that 'RBC' should be a near instantaneous carrier of emotional messages; and that organisms can easily switch from Rhythm Base to Signal Base. This is as we have just seen for the north-south rhythmic messages, versus our east-west non rhythmic communications."

"Explain again how the non-rhythmic parts work, s'il vous plait," comments Elliott.

"The main opening rhythm this evening was one short sound transmission every 60 seconds; and the four windows were each four seconds wide centered at 12, 3, 6, and 9 on our computer clock or on your watch face. When most of the whale blows and tail slaps occurred during these 16 seconds each minute, we interpreted our results as 'RBC.' But if they happened outside of these 'duraTions,' that is, during the other 44 seconds each minute, then we interpreted the results as 'SBC.' Confirmation came when all affirmative and negative messages were placed in either of two, four second durations, the first being the 'onTime' window. The second, was the 'offTime' window, which followed the first by half a cycle or what we call half phase."

"How do we know that Ida was telling the truth?" comes a stinging response from Elliott.

"Good question, and now we'll have to merge the scientific theory of 'Rhythm Based Communication' or the low stress altruistic aspect of Nature with our philosophic thoughts about what it is that makes the biosphere the way that it is. Alex, myself and many others believe that inherent trust, honesty and altruism have evolved into and are associated with 'RBC' in contrast to the inherent distrust, dishonesty and 'survival of the fittest,' that have evolved into and are associated with 'SBC.' The crux of the debate might migrate towards the contention that evolutionary theory by natural selection seems correct but not complete."

Alex then joins the discussion in a dynamic fashion, one that I have only seen in his letters, not as yet in his verbal communications. "A traditional and logical statement in evolutionary biology, all the way from Charles Darwin in 1859 to Dawkins and others of today is: 'a selfish mutant would quickly out reproduce its altruistic competitors.' But we believe this to be only true for high stress, 'fight or flight' situations in Nature, which we claim are 'SBC' conditions and may not occupy the entire time and space continuum of the Earth's biosphere. This conjecture is based firmly on the evidence that if an organism changes its altruistic state and generates 'destructive' feelings, harmful concepts, or a contrary, non-rhythmic mental state, accordingly this information is transmitted by natural processes, such as the observation of 'Conventional time.'

Consequently, such neighboring organisms rapidly switch to 'SBC,' 'survival of the fittest' and the true evolutionary theory of natural selection."

"And how do you tell which communication the animal is in?" asks George.

"For Rhythm Based research we use the computerized traffic light for stress, which indicates the changeover from 'SBC' to 'RBC' and vice versa. In Nature, information in the form of conscious or unconscious concepts, as well as traffic light feelings that are possibly only in the unconscious, are mechanisms to move organisms into and out of an altruistic state. The ethics of altruism is honesty; its environment is low biological stress."

"Is there anything to prevent 'RBC' from becoming misused?" asks George.

"Yes, and the answer is very powerful, very clear and absolutely essential to the survival of all ecosystems of Earth. Recall that 'RBC' requires and produces low biological stress. Also recall that 'RBC' is a rapid carrier of emotional messages and that organisms can easily and quickly switch communication from a pure Rhythm Base to a pure Signal Base. Consider the following:

Scenario one. Organism A meets organism B and there is no fear, no externally caused stress by the presence of the other organism or an altered environment, and after 'synchronizaTion' there is a 'passkey invitation' by A to communicate with B using our 'RBC' or 'Rhythm Based CommunicaTion.' We believe that each of these two steps: a, 'SynchronizaTion' and b, Passkey Message Mimicry, can lower the biological stress of both A and B whereby they may remain in 'RBC,' where trust grows and altruism is a naturally developed code of ethics. Each organism stands to gain in biological survival potential, not only because of the lower stress but also because of an altruistic code which may be intimately connected to a powerful and rapid form of rhythmic 'communicaTion'. The complementary, new and additional component to biological behaviour, which is illustrated by scenario one, cannot exist without the following conjecture."

Scenario two. Organism A meets organism B and there is fear, and externally caused stress by the presence of the other organism or an altered environment. The evolutionary theory of natural selection prevails and communications are entirely encoded in, signals, signs and symbols. But -

Scenario three. Organism A meets organism B and there is no fear, no externally caused stress by the presence of the other organism or an altered environment and after 'synchronizaTion' there is a passkey invitation by A to converse with B using 'Rhythm Based CommunicaTion.' So far this is the same as in the first scenario. Now, organism B changes its altruistic state and generates 'destructive' feelings, harmful concepts, or a contrary non-rhythmic mental state. This new information is transmitted by natural processes, such as the observation of 'Rhythm Based Time' windows, to organism A which instantly switches back to 'SBC,' 'fight or flight,' 'survival of the fittest' and the second scenario."

"We have seen this happen, mainly for the internal stress of hunger, a category of 'destructive' feelings, whereby a whale drops the 'Rhythm Based CommunicaTion' and our computerized traffic light, measuring stress, switches from green to red before the ensuing feeding commences. It is now possible to measure the difference between the altruistic, 'RBC' state and the 'natural selection,' higher stress state of 'SBC,' by analysis of signal timing. These appear to be two distinctive rooms in the 'House of Nature' and organisms may move freely between them. Both such compatible, 'living' rooms support strong, complementary evolutionary selection pressures for survival."

The vehicle slows to half speed as it is dusk and 'Moose Alley' lies around the next corner! Soon this is considered a wise move as three cars ahead are on the roadside while a giant bull moose with two mother-calf pairs are sauntering up to the pavement on their way to the scenic drinking brook on the lower south side. We pull alongside as George loads his fastest film. Now, five magnificent animals are standing almost motionless on the road, gazing toward us, then three more juveniles join and we have yet another 'Moose Alley Roadblock!' A well dressed senior lady and a younger teenage lad, leave their car and seem to be considering a move toward the animals, when Nick motions them to 'freeze' next to their vehicle. Using our customary moose rhythm of one signal every 32 seconds, the field computer indicates high animal stress as we transmit short, sharp, muffled whistles and observe all of the animals to turn our way, ideal for George's portrait photography. In three rhythmic cycles the bull moose twitches one ear in synchronization with our whistle signal, and Nick changes over the program to passkey message mimicry. The stress light soon goes to green and we slowly leave the car and invite the others to come toward the animals to view the communications experiment. Both mothers move ahead to the drinking water but the bull and the juveniles, seemingly inquisitive, maintain their pavement places. Following a reciprocal greeting, Nick leads the lad slowly over to the nearest, young moose and both, each in turn, stroke gently with their knuckles, just below the animal's eye. Nick continues the communications research with a 'late, onTime' message for each of several, designated, boy-moose contact sessions and surprisingly receives a mimicked message from another juvenile. The large male, without moving, has apparently handed over communications leadership to another, perhaps an opportunity to switch our statement to a question!

"Try the interrogative Nick"

"What's up Doc?" says Elliott.

"Nick has established, by conditional response, a declarative message for physical contact, and we think that slight jealousy might be creeping in with the other juveniles."

By making a simple 'late-early' exchange, Nick transmits the message 'early, onTime' and forthwith the animal that had previously mimicked the 'contact message' gives an 'onTime' double ear twitch, a clear indication of the affirmative response. Nick slowly guides the lad, who has introduced himself as Otto, to contact the second juvenile, and motions Elliott to maintain, with care, the original human to moose, soft whistle signaling. The bull watches intently, moving slightly closer; now we can estimate the greater than one meter width of his antlers.

Two smaller children leave another car and are cautioned to stay back along with two adults from the same vehicle. Otto, leading these amazing contact encounters, then moves on to a third juvenile while Elliott moves slowly to the second and Alex joins him. George calmly suggests that everyone shift to the left slightly in order to align themselves with the giant antlers, for another memorable photo opportunity. Hans maintains traffic control!

"Otto! Look behind you!"

Both calves are returning up the bank and approaching the lad. With one hand for each, Otto lovingly strokes them while using his best German language to softly express his feelings.

"These animals don't understand language Otto, you must express your feelings with rhythms, slow movements coupled with the friendly tone of your voice."

His nearby grandmother translates briefly although I feel that the lad understands, by natural instincts. The two female moose approach cautiously, slowly. The three juveniles saunter to the brook side while the younger children are motioned closer to the friendly calves.

"Where are you staying Otto?"

"At The Village Inn in Trinity, we just arrived," replies his grandmother.

"That is where we are heading. Did you meet my wife, Chris?"

"Wow! You must be Peter! Chris sent us up here to see a moose! Little did we know that all of these animals would be watching us! This is a thrill of a lifetime for Otto," replies Katrina, his grandmother.

The bull moose slowly saunters to the brook as the juveniles begin to return. There are now a few additional local people, nearby but staying in their newly arrived cars; some look inquisitive.

"It is safe to let your youngsters closer to the young calves now. Are you their parents or guardians?"

"Thanks, their names are Simon and Sarah and we are indeed their parents, Joan and Stephen Simpson from California," replies the father.

The children enter Nature's petting park and Dad takes a 'zillion' pictures. George takes a few while softly saying: "Yes, a picture is indeed worth a thousand words."

"How was your flight across the 'pond' Katrina? I certainly knew about your 'whale contact' expedition, starting tomorrow."

"Excellent," she replies. "It seemed to take less time than our comprehensive tour of Beethoven's birth place!"

"This adventure vacation is certainly a fine present for you to give Otto and we seem to have started it with what appears to be a five star 'animal contact' experience."

More cars arrive. The moose roadblock is holding solidly but a few locals now look both inquisitive and a bit impatient. A few cars back is a prominent fisherman from Melrose, respectfully called 'Uncle Tom' by everyone from youngsters to esteemed seniors. He knows of our research. When Tom exits his driver's seat he quickly gathers some inquisitive 'locals' like a planned football huddle with a popular quarterback.

The bull moose returns. The curtain suddenly descends on our drama as all eight, outstanding animals leave for their higher grounds with slight haste, as the huddle disperses. Otto rushes over to give Nick and I big hugs, followed by tears of joy in his eyes. He exclaims that nowhere in Europe could this have possibly happened. I reply that he should just wait for his coming 'whale contact' expeditions, perhaps even by early tomorrow!

CHAPTER 10

Bonavista

The 6:00 a.m. forecast calls for moderate northwest winds and cloudless skies, an excellent situation to be with whales along the southern shores of towering Cape Bonavista. All is quiet on the home front except for the sounds of perking coffee and a muffled radio broadcasting weather.

This morning there is a detailed email from friend David about the unraveling of some of the mysteries of superluminal (faster than light speed) velocities, and I am wishing that what William and Alex know could soon be learned by experimenters endeavouring to study individual motions of light particles or photons. I reply to David in detail that there are indeed two independent components of time (which he has been pondering now for some time) and therefore, there must be two independent forms of space! After all, space can be defined by the measured time that it takes an amount of information, either particle or rhythm, to travel at its given velocity. Suppose the particle is a car traveling in a straight path at 100 km per hour. Then its space traversed is defined by the measured time (perhaps as recorded from your watch), say one hour of conventional time t, multiplied by the average velocity of the car. This then becomes one hour multiplied by 100 km per hour giving the spacial car displacement of 100 km. The information, which in this case is dependent on the mass and energy of the car, has moved so as to define what is indeed a 'physical space.' And let us presume that no mass or no energy can exceed the speed of light so that physical space is determined by physical properties as logically established by velocity limits. Just look at someone who is a meter away from you, and you can completely comprehend what distance, in our physical space, is indeed a meter! There is something, however, that can travel between you, faster than the velocity of light, and that is information encoded in a second and independent temporal form! And we label this 'informaTion.'

Send such a person a message, perhaps a feeling or an emotion, that is partially or completely encoded in 'Rhythm Based Time' ('orthogonal time'), which biological variable presumably can be 'rotated' by the receiving person, or organism, at an angular velocity faster than the velocity of light. This is a consequence of there being no mass and no energy directly involved in such a 'rotation.' Then a second form of space emerges! In the case where the two organisms have synchronized biological clocks, and thus an information component encoded in 'Perceptions of Lateness Relative to the OnTimeness' ('orthogonal time') of these two clocks can indeed be superluminal, in the mind of the receiver. This new form of space is a 'communicaTion' space rather than a physical space. Perhaps, because the two different forms of space are often combined in common messages, we have not previously recognized their individual differences. The former 'Rhythm Based CommunicaTion' Space (or 'RBC Space') can be defined as above, by the measured 'Rhythm Based Time,' in the mind of the receiver, that it takes for an amount of 'informaTion' to travel directly from one person to another, at its given conventional velocity, which, let us presume, may not exceed the velocity of light. But the internal rotation, of the receiver's clock, presumably may exceed the velocity of light (because no mass and no energy is directly involved in 'RBT' clock rotation, and in this case the 'RBC Space' becomes smaller than the physical space, and the 'informaTion' that is encoded in 'orthogonal time' (not in mass or energy) travels superluminally. This reality appears to account for most apparent paranormal phenomenon, such as that of two people receiving the same feelings at the same 'Time,' seemingly independent of their physical space. It possibly accounts for newly acquired experimental evidence of superluminal velocities. It is undoubtedly a new window into future communications research.

At the flick of a mouse click these new concepts travel to David in seconds and as Marshall McLuhan's philosophy of an emerging 'Global Village' has partially arrived, the same mail may similarly be almost instantly sent throughout the global 'web world.' Some however, like myself, are also fond of curling up with a good book by a warm crackling fire, surrounded by calm silence, or better, by a quiet rendition of Bach's 'Ave Maria.'

Alex arrives from his suite located on the same floor but in a quiet corner of the Inn; Elliott arrives from across the hall.

"I've an idea about Time Rotation," says Alex, as we prepare ourselves for a new Russian knowledge 'knockout!'

"And I've an inquiry about who's singing one of my Bach favourites," adds Elliott.

"That's Andrea Bocelli and now that you are both here I suppose we should follow it with Alex's favourite Tschaikowsky Concerto."

As the music plays and a pre breakfast appears, Alex and I listen to the unfolding story of why such an amazing man, now in our company, could qualify for the highest world recognition.

Elliott is fascinated with Nature, with human communications, with animals and particularly with our discoveries in human-whale/beaver/fox/eagle communications research. He is a great listener to novel ideas and always, in my recollection, makes definitive comments. He likes our ideas and progress with adventure expeditions, now easily converted into 'Whale-Contact' or 'Whales Watching Us,' and he is awe inspired by the potential universality of 'Rhythm Based CommunicaTion.' He loves his children, Nature, and in particular, our Canadian beaver[22] and he initiated our terrestrial mammal communications research by proposing a similar methodology, using flashing lights instead of underwater sounds.

Elliott proposed that beaver communications research should become as independent of language as possible, a positive unifying concept for Canada.

---

[22] Canadian beaver (Castor canadensis)

Now students can use the internet to compare 'Nature's concepts,' which are more modest, more truthful, more universal, than language. His vision may not only unite the country in many ways but also aid conservation, potentially create a negotiated peace between all animals and eventually, unify nations throughout the biosphere. He envisions the 21st century as not a continuation of humans involved with 'Nature Viewing,' but rather a universal involvement with 'Nature-Contact.'

Then Elliott slips out so that he can meet fishermen and fisherwomen at the main wharf, preparing for a busy day with a now highly varied number of species, including lobster, snow crab, sea urchins and more.

"I feel that if any of us receives recognition for insight and discovery that we should donate it to a Foundation for the future study of 'Nature-Contact' worldwide," says Alex.

"Here's a computer copy of a letter sent this morning explaining our views on paranormal communications phenomena," as I hand it to Alex. "And previously I have explained to David our feeling that there are no extra senses needed in communication theory, just an extra elementary comprehension of our sense of biological time. Would you care for some fresh coffee?"

"Fine my friend," as he reads and talks simultaneously. "My idea this morning is that the receiving clocks for such communication must be double helixes such as are in every living cell and that the 'rotation' that you refer to in the letter, could be caused by the shunting of energy across the rungs of a DNA 'twisted ladder,' until a late signal within one of the ladder's uprights, has time to catch up to the 'onTime' signal in the other protein chain," comments Alex.

"In that case what we need is the ladder's information wave velocity as well as the time required for a so called delay cycle across one rung, the nucleotides, and returning on another. This should tell us just how much faster than the velocity of light a visual signal could be comprehended by a receiving organism, using what is essentially a 'Rhythm Based Time Rotation'."

Otto walks in from his room next to our now noisy kitchen as the computer is printing out additional morning emails.

"Good morning," I say in my broken German, followed by French.

"Good morning to you both," he replies, in all three languages!

Alex, who speaks numerous languages, asks the lad if he had enjoyed pleasant moose dreams, and a giant smile appears with no linguistic explanation necessary. Alex gives the slightly shivering, sleepy eyed boy a giant Russian bear hug and I hand him a large glass of fresh orange juice, which the lad had fondly eyed upon arrival. Otto locks his eyes onto a large kitchen photograph depicting a humpback leaping clear of the water beside towering cliffs. There are astonished faces in the background of the picture, all safely aboard the *Ceres*.

"Buckelval?" and he repeats in English, "Humpback? Why do you use the word humpback?"

"Because it has a hump on its back, the size of a Canadian football, just ahead of the dorsal fin and easy to recognize," shouts his grandmother from next door, first in English and then translated so that Otto can understand every detail.

"Wow! How did you remember that?" I ask as she too appears in a Village Inn bath robe.

"I read it last night in your book 'Dancing With Whales' Peter," she replies.

"Juice, coffee, tea?"

"Yes and danke shane, thank you. I would certainly enjoy some tea, just like your British guests prefer, as for example the nice family that is, next door to us," she replies.

"It's nearly student seminar time; we should get 'On the Go' as our Newfoundland fishers would say!" The time has come for 'Whale Contact' clothing as I give the day plan to Nick.

"Nick you go ahead and take your mother, Otto and his grandmother, Katrina, to the *Ceres* and we'll plan to land at the Bonavista airstrip at 10 a.m. sharp, with radio contact. Kirk will take other folks out from Trinity here. The Northern Peninsula black bear team should be prepared to head out tomorrow and the beaver team should head up to Notre Dame Bay at about the same time. All preparations should be nearly complete, except, be sure to check the software analysis techniques with Tim before departure and order food supplies from Chris. The cooks are making masses of date squares and cookies, hopefully to last a week or more."

Enter Charles asking, "What do we do about fuel at the camp site near Sandy Harbour?"

"Stan has stored extra at his lodge, which is only a few kilometers away, we should bring emergency fuel on Ceres and you are mainly studying black bears not whales, blue, gray or otherwise!" I reply.

"How are you getting to Bonavista, Dad?" asks Nick.

"We'll take the five passenger Beaver so that we can locate the humpback 'heard' before landing. Everyone, have a productive day."

Hans is back from jogging, discreetly claiming a new record to and from the summit, and having seen hundreds and perhaps thousands of dolphins just outside Trinity harbour.

"The Newfoundland language would describe the dolphins as: 'Tousands and tousands, there must have been a hundred!', with their word, tousands literally meaning many," I remark.

"So are there three languages in Canada: French, English and Newfoundland?" asks Hans.

"Exactly, and others besides. We now have tens of thousands of words in a Newfoundland dictionary. For example 'me son' means my friend, not necessarily a relation!"

"Gentlemen!" I announce during breakfast, "We have a convertible sea and pavement landing plane arriving in half an hour at Jack's wharf so that we should be able to photograph whales from the air before landing near Bonavista."

"Incredible! adds George. "Can we keep the window open for photography?"

"The starboard, middle passenger-seat side-panel completely removes for this, George, but we all have to keep our seat belts on permanently!"

"How much storage space and what plane type?" asks Elliott.

"It's a Beaver, and if you place any boat gear in the Explorer, Nick will move it so as to meet us at the landing site, near where the whales were feeding last evening."

"Who wants good, old, Canadian maple syrup for their waffles? Mr. Elliott, are you feeling Canadian today?" inquires Penny jestingly!

"Every day of my life Penny, and every night of your life!" replies Elliott.

"Every night but some sir!" as all look on astonishingly. "As how do you know that I wasn't born before Newfoundland joined Canada?" says our star waitress, amidst relieving sounds of laughter and pure, 'NowTIME' joy.

"I wouldn't mind being your knight, Penny," comes a reply from Hans.

I think that Elliott contemplates bundling Penny up to Ontario and then recommending her for a first class urban restaurant. Sometimes Ontario could do with a lighthearted dose of Newfoundland wit!

The buzzing noise of the plane circling over Trinity marks pilot Dale's imminent arrival. And Jason has a smaller Zodiac ready at Jack's wharf to facilitate our departure. Following a picture perfect landing on the southern of the three harbour arms, the shape of which suggested to the early Portuguese the name for Trinity, I radio to the approaching Beaver.

"Peter to Dale, shall we bring you some fresh coffee, over?"

"No thanks 'me son,' we'll be airborne when you're ready," comes the reply.

"Lets 'Load and Go' gentlemen," and then Jason skippers us a stone's throw to the plane.

Takeoff is noisy but routine, and in a brief moment I see a virtual sea of dolphins, a species called 'jumpers' by the local fishers, 'lags,' or white beaked dolphins by scientists worldwide. Dale knows to circle them on the right side for our photography and I expect that he will be able to stall the plane down to about 100 meters and make George a very happy passenger.

"Dolphin 'heard' at one o'clock, range half a mile, we'll circle clockwise," I announce.

The plane quietly drops half way to the calm seas and what a show the animals perform!

"Notice the herring school trapped between the two forward arms of the dolphins."

The Beaver rises to attempt a second circle and George's camera clicks in bursts of many frames per second. Certain 'lags,' while spinning rapidly, jump two to three body lengths into the air, and then drop on their backs with enormous splashes, presumably to herd the herring.

"Lags? What is the semantics?" asks Alex.

"It's an abbreviation of the Latin for ocean dolphins. The full name is white beaked dolphins, *Lagenorenchus alborostris*. These are large animals, many hundreds of kilograms per adult."

"Amazing scenery on the left side," comments Hans as we leave the lags and move eastward along towering cliffs and numerous seabird colonies.

"Eleven o'clock, about a mile ahead, five or six fin whales in close formation," I announce, after Dale had pointed to them with one of his invaluable, flight controlling hands.

"Let's swing in and circle them for a surfacing sequence. It's shallow water so I expect them up in about two to three minutes," I say vociferously to Dale.

We swing over the Green Bay Barrens and there on a grassy meadow are two red fox adults playing with their rather large 'handful' of young kits, two black and five red! Hans notifies Elliott but they are on the wrong side of the plane for George and Alex.

"Blows at 3 o'clock, half a mile, heading east, six large fin whales swimming as if they're one."

We circle twice this time, stalling down to only a few whale lengths of the animals! We can all see the bright white, lower right jaws that distinguish the species, as well as the half length juvenile, almost 'cuddled' by the adults.

"That's Oscar's family," I announce. "He is perhaps the largest animal in the Northern Hemisphere, as most older fin and blue whales were taken by the industry earlier in the century. They appear headed to Bonavista!"

"Do you think Oscar is longer than Alexi the blue whale that we rescued?" asks Hans.

"Yes, just slightly, and much older as well."

We pass the spectacular light house at Catalina and then head onwards to the puffin, gull and whale feeding grounds off Elliston.

"Dale! That looks like a large yacht, apparently aground on Flower's Shoal," as I point ahead and slightly to our right.

A flare fires upward and across our flight path. We circle to see arm waving crew members, some in a small inflatable, others apparently handling luffing sails, while ocean breakers render the decks awash.

"I've got them on channel 16." I say to Dale as we complete the second circling.

"Beaver to yacht, come in please, over."

"Yacht *Atlantic Explorer* back. Can you send a rescue and towing boat A.S.A.P, over?"

"Roger, rescue craft expected in 15-20 minutes. Stay calm, over."

"10-4, *Atlantic Explorer* standing by."

And we turn toward Elliston harbour while I switch to channel 9.

"Peter to Nick. How do you read? over."

"Weak with static Dad, over," replies the ever alert Nicholas.

"Mayday alarm! Grounded yacht on Flowers shoal. Can you leave for rescue, over?"

"Roger. I read to rescue a yacht on Flowers Shoal. Leaving in a few minutes, standing bye on 9 and 16, over," replies Nick.

"Roger. Switching to yacht *Atlantic Explorer* on 16, out."

"Beaver to *Atlantic Explorer*. Rescue craft *Ceres* arriving shortly, over."

"Thank you Beaver! standing by," replies a relieved sounding and probably very wet crew member.

Dale stays high over a famous bird Sanctuary, 'The Gull Islands,' off Elliston, as enormous flocks of puffins swirl about in their 'Dance of Navigation,' their seeming ability to all turn together, and as we have discovered, their doing so without apparent signals from their leaders. This dance, which we have similarly studied from schooling fish, is perhaps the strongest known evidence for 'Rhythm Based CommunicaTion,' whereby the navigational information is encoded in 'Time' instead of in signals, as it most often is for humans.

"Ida's 'heard' is at 2 o'clock, half a mile, with at least four humpbacks," I announce.

Dale knows to circle right as these identification photos are a main reason for our flight, although arranging a rescue at sea has obviously attained center stage for the moment.

"There are minke whales amongst the humpbacks. Can you see and/or photograph the white armbands which distinguish this species, George?"

Round and round we go, at levels low enough to see the eyes that are presumably watching us, and indeed until the whales have become accustomed to the modest engine droning. Then just after we have made the move to leave, Ida breaches!

"Wow!" shouts George. "Can we try for a closer breaching photograph?"

Dale nods affirmatively, apparently there is lots of fuel left. He circles back and as I try to estimate the possible breach timing he manages to place George almost beside and eye-to-eye with the enormous leaping (and possibly laughing) 'Great Whale - Ida.'

We notice Nick racing by with apparently two Bonavista fishermen.

"Can you rise up over that western cliff again Gene and I'll monitor the distressed yacht?"

"*Ceres* to *Atlantic Explorer*, over," echoes Nick's radio voice.

"Roger *Ceres*, over."

"You should just see us coming in from your northwest, over."

"Roger, we have a female in severe shock, two broken limbs and a total of seven crew members, over."

"Beaver here. We will have medical crews sent to the Elliston main dock, over."

"Roger Dad, stand by one, over."

"Let's head down to the Bonavista airstrip Gene, we'll be finished flying for today," I say.

We land comfortably and the taxi which we had called awaits to take us all to the Elliston wharf. Nicholas arrives to three waiting ambulances and additional splints are administered aboard *Ceres*. One younger lad seems to be unconscious. The captain, a fit and seafaring man, insists on helping with the rescue of the yacht so he aids the others into vehicles and sends them off to the very excellent Bonavista hospital nearby. He shows somewhat moist eyes, but with dramatic signs of relief he is able to discuss the accident and a proposed plan. I arrange with our four air passengers plus Chris, Katrina and Otto (who obtained a ride with friends), that this morning's coffee break can be on the grassy meadow across from the local store. "Nick, the captain, two fishermen and I should be able to return here before too long and we'll keep in touch by radio." Wally Newton appears and says that he has contacted a 'longliner type' fishing boat offshore named *Sue and Sally* that will assist if we need her. I thank him sincerely.

"Come on captain, let's see what we can save," and we are off at full speed in the *Ceres*, to the famous (locally that is) breakers of Flowers Shoal.

Nick attaches our two centimeter nylon towing rope to a specially designed bridle, through two stem bolts and then it is stress braced forward to the base of our aluminum mast. We depart Elliston and soon we are nearing the *Atlantic Explorer*.

"There's your inflatable tender, captain, let's place it across our bow as it might come in handy later."

A nod of approval allows a slight course change of the speeding *Ceres*, to the empty tender. Shortly afterwards we survey the damage to *Atlantic Explorer.*

"Captain, how about if Nick retrieves that flapping foresail? You've done a fine job of furling the mainsail"

"Ok, but could he first retrieve some precious things which I hope are still in the drawers under the chart table?" asks the captain.

"Go for it Nick but watch out for breaking waves, loose rigging and the swinging main boom."

"*Ceres* to *Sue and Sally*, over."

"Peter 'me son' do you need a larger boat to help out?"

"That's you Gerald. Please come and tow this handsome yacht to Elliston if you can help. I'll rig the lines and we'll be ready when you get here."

"Done, 'me son' but gives me twenty fathoms of tow rope so's I be clear of the shoal," replies captain Gerald.

The foresail is now aboard *Ceres,* along with loose deck materials such as a spinnaker pole and cushions and the *Explorer* captain looks more relieved with his several boxes of valuables safely retrieved.

"I'll throw you the tow line Nick. Tie a bowline around the mast, then lead it forward through the bow chalk. Lash the tiller amidships. Here's two suitable lines, and then I'll come closer to get you a.s.a.p."

Nick is soaked from decks awash with foam, but he manages to prepare the *Atlantic Explorer* and then leap back aboard with us for the trial towing.

"How are we going to get her off?" questions the captain.

"Just look aft mate, we have 200 horsepower of four stroke muscle and if we can rescue 100,000 pound whales we should be able to get you off, provided Nick, that we don't wrap the tow line!"

Slowly we manoeuver eastward as Gerald approaches. The tow line tightens and stretches, as designed. Neither vessel progresses forward. Half speed! Slowly faster! Nothing changes!

"Dad! Give her a jolt just as the ground swell heightens, when I say," suggests Nick.

5-4-3-2-1-Now," repeatedly counts Nick, as he adjusts for the rhythm of the larger swells.

"She moved" the captain shouts excitedly. "She moved again! And again! We're free!"

"Gerald, she's clear, we'll throw you the towline when you're ready."

"My god, she's sinking! What can we do? She'll go to the bottom!" cries the captain.

"Gerald, can you guys rig a pump and tow her in alongside?" I say by radio, possibly unheard by the captain.

"Easy, 'me son' bring her over soon as I gets me fenders rigged port side." replies Gerald.

"Peter, she'll go to the bottom in ten minutes," says a partially shocked captain.

"It's only six feet deep here, so not to worry captain. Your engine is still high and dry," replies Nick.

The whir of an electric generator starts and as we maneuver the *Atlantic Explorer* next to *Sue and Sally* the snake-like feature of a two inch bilge pump winds its way into the flooded bilge of the yacht and a baffled looking captain begins to beam with the joy of a man just saved from depths of despair.

"Why don't you ride in with your vessel captain, and Nick and I will try to retrieve anything floating, including those cockpit cushions and paddles blown in by the shore."

"Ok and thanks a million to you all," replies the captain.

Shortly we are back at the Elliston wharf loading our research crew, including Chris, Katrina and Otto.

"Lets avoid those reporters and head right out for our new highest priority, 'Whale-Contact' mission."

We off-load the rescued material, on-load our research crew, plus dry clothing for Nick, and are ocean bound momentarily. Ida, Andrew, Hubert, Cecile and their companion humpbacks are in relatively calm, shallow waters just south of The Gull Islands. As Gerald passes us we wave, as he slowly guides Atlantic Explorer to the sheltered wharf at Elliston.

"It looks as if that rescue story will end reasonably well for the *Atlantic Endeavor*," comments Elliott, "so that by reverse logic we can conclude that 'all is well,' hopefully, for them!"

"That is provided that they can easily repair the hull leak. But in and around Elliston, from Trinity to Bonavista, we have some of the finest ship builders on the island, and facilities to haul her out in short order."

"Blow, 12 o'clock, 300 meters, heading our way, two humpbacks, 'en echelon;' second blow in 10 seconds or so," announces Nick.

"Start a 1 minute 'alpha rhythm.' I'm cutting starboard engine at computer clock 'onTime'."

"What does 'en echelon' signify?" asks Alex.

"The local folks would say 'side by each' with one slightly behind the other, for speed. It comes from the geological expression for glacial drumlins and is similar to geese flying, in order to reduce drag forces on a trailing animal," I answer.

"5-4-3-2-1-mark, (and an engine goes silent). Blow, 12 o'clock, 200 meters, its Hubert leading Cecile," loudly comments Nick.

"During the next longer feeding dive, Hubert will force the small, capelin fish toward Cecile, using his very long white flippers, or arms, and then surface 'en echelon,' but ahead."

Very slowly I reduce speed on the port engine so as not to produce an ill timed or misguided, 'high risetime' 'RBC' signal.

"Ida and Andrew are up by South Gull Island," observes Nick.

"There are three or more large whales together, blowing now on our port side," exclaims Hans.

"Bearing and range for Nick's computer entries please, Hans."

"Tenish and a mile I guess, perhaps closer to two miles!" replies Hans.

Nick, higher above on the mast, taps me on the back with his toe as a 'ten second warning.' This is a reminder that I must stop momentarily, contemplating what could be in the conscious minds of Hubert and Cecile and switch to a different 'NowTIME,' such as what would be an appropriate, low stress and perhaps familiar signal to transmit to all the nearby whales.

"5-4-3-2-1-mark, 'alpha one,' you were half a second late!" shouts Nick!

"Sorry old buddy!" as I retrieve a tethered sound pinger after its two seconds in the water.

"George! 12:30, white under water, coming up at about 80 meters in five seconds," calmly says Nick. "Blows, Cecile leading Hubert, sounding dives, two 'fluke overs,' watch for the flippers moving out like airplane wings just before the sounding!"

A sight to behold! Two giant tails smoothly lift into the sky and send the powerful, 'Signal Based Message' to any and all of Nature's watching eyes: 'We two are going down deep.'

"Notice gentlemen that there was no sudden motion, or even what we refer to as a 'high risetime signal,' and thus we have no synchronization, no common rhythm between Nick's computer clock and any whale, and no 'Rhythm Based CommunicaTion,' yet!"

Kick tap, countdown and 'alpha two' follow as we gently slow and turn the *Ceres*. Just off our bow, dozens of harp seals pop up their canine type heads together; everywhere there are puffins zooming by in small flocks, at about thirty knots! The sea is alive!

"Those seals are certainly synchronized!" comments Elliott as they all dive in unison.

"Ah, but that is encrypted 'synchronizaTion,' where timing is very probably known only to the species. The seal's 'alpha time,' or time between the rhythmic cycles that govern necessary navigational information, is one of their 'passkeys.' Evolution should make this 'Time,' as well as additional seal 'passkeys,' extremely difficult to penetrate unless the animals voluntarily give such information to friendly altruists."

"5-4-3-2-1-mark, blow, 1:30, 'synchronizaTion' with Cecile, then comes Hubert circling right. I think that they recognize either the *Ceres*, our alpha rhythm, or both." exclaims Nick, rather more excited than before.

"I experienced a sudden chest feeling, somewhat joyful, when Nick hollered blow, and I now instinctively seemed to know where to look!" contributes Alex.

"Same here, I also got that feeling!" says Otto!

"Yes indeed. I can confirm that gentlemen! We have 'synchronizaTion'," I reply.

"That's amazing! William ought to be here. Simultaneous emotional energy was just transmitted, over 50 meters, between a cetacean and us, at some unknown, very high velocity," adds Alex. "And it apparently wasn't either visual, sound/touch, or chemoreception."

Otto initiates inquiry to Katrina but Alex sensing the nature of his communication, immediately responds, "Chemoreception involves basically the senses of taste and smell, Otto." Cecile swims to and then under the *Ceres* and slowing to much less than human walking speed, she spreads wide her wing-like flippers, now outstretched larger than the Beaver plane that brought the six of us to Bonavista. She uses them as ailerons to control her lateral balance.

"Now Otto," as he stretches away out over the forward, starboard inflatable pontoon, "you are being watched by a whale. This is what we mean by Nature-Contact."

"Next time grab the glass bottom bucket in the aft port locker," Nicholas calls down from his mast perch above. "Place it just underwater and get your eyes as close to the glass as possible. Then you'll get eye to eye contact, hold on, here comes Hubert; grab it Hans, there behind you."

Otto is given the bucket, then like a perfect gentleman, hands it to Katrina and physically but gently coaxes her to what she describes as: "The largest eye I've ever seen, and Hubert was watching me! It was astounding; one of the very greatest experiences of a lifetime!"

Kick tap, countdown and 'gamma one' follow as we simultaneously stop all power, and I humbly request for a totally 'silent ship.'

Now speaking ever so quietly, I declare, "expecting a reciprocal overlapping greeting from Cecile on coming countdown; wide angle lenses everyone."

George, standing high in front of the console, rapidly detaches his enormous 'zoom' lens and fastens his close-up camera lens just as Nick's hand held, 'conductor clock' alarms us with a ten second warning beep.

"5-4-3-2-1-mark. Holy Smoke!"

Cecil vertically rises half of her length, with dorsal side facing the *Ceres* so that both eyes can probably see us all! She is at a range of no more than half of her length and she does not exhale, otherwise we would all be soaked from the several liters of moisture in a humpback blow. George snaps off half a roll of film, full frame at the apex of Cecil's spyhopping manoeuver! She slips in silence back to depths.

"Wow," says Chris.

Kick-tap, countdown, an Alpha Signal and nothing happens from either humpback, as expected.

"What do you predict now?" asks Elliott.

"I know not the 'WHAT,' my friend, but the 'WHEN' should be in about ten seconds."

"Coming up, two whales, white under water, 3:30, 50 meters, now!" shouts Nick.

"Synchronized blows, water coming, shield your cameras."

"The Stress Light has lowered to green, the 'RBC' level. A Massage Mimicry Light is flashing. A-1 Dad!" says Nick.

Katrina has tears of joy, Alex stands impressed, with mouth half open while Otto searches underwater with his head far down inside the glass bottomed bucket!

"In computer language you could call that a 'handshake,' but it was more. It was a 'reciprocal overlapping greeting' which could be considered almost akin to a 'Russian bear hug!' For example, if a friend, at a distance, takes her hat off before you have replaced yours, then her greeting is 'overlapping' in time, representing a response with previous familiarity."

"A handshake with a whale, with Nature, that's an interesting thought Doc." says Otto.

"The boy has a rather penetrating mind," comments Elliott to Hans.

"Nick, please program interrogative of breach."

"Are you asking the whales if they want to jump?" asks Alex.

"Affirmative, but the outgoing message, encoded in 'Time' only, will take several minutes. These whales already know a crucial step in our learned 'Rhythm Based CommunicaTion,' and that is how to let an arbitrary variety of signals compose the timed 'RBT' messages."

During the first computer conducted window of transmission, both engines are turned on simultaneously. During the next window, hull mounted acoustic transmitters are activated by console buttons whereby we end the message in unison, by sending out six sounds. This is a cooperative effort between Alex, Elliott and Hans tossing over tethered 'pingers' and Otto activating the console switches including a depth sounder, all for approximately two seconds but commencing exactly on Nick's command.

"Synchronous blows at 12 o'clock," shouts Nicholas excitedly.

"Were they in the 'onTime' window Nick; is the 'Affirmative Light' flashing?"

"The humpbacks have just answered 'Yes' to our question regarding breaching, or jumping so I'm expecting two things: they are in a non feeding, playful mood and you should get 'onTime' breaches every 60 seconds, - - and starting right - - now!"

Ida jumps skyward just a few hundred meters beyond Hubert's last location. She lands with an enormous splash. Just a minute later Cecil and Hubert perform synchronous breaches, landing so as to produce one powerful sound, which undoubtedly travels throughout this humpback 'heard.' George zooms out, then in, then out again, to capture for our memories just as much as possible. Katrina is seated in silent wonder, Otto is half way up the mast with Nick. Another minute later, Ida and Andrew perform closer synchronous breaches, then a fifth whale, a juvenile, joins in and with a playful display leaps entirely clear of the water so that the nearby shoreline can be seen beneath the horizontally 'flying' humpback. Cars stop, people gather, the sea is alive!

Dolphins approach. These are hundreds of white sided dolphins[23], with their golden diagonal blazes glistening in the sun. 'Spinners' or North Atlantic white beaked dolphins[24] also arrive. These are a larger species that often rotates two and a half or even three and a half turns while airborne, splashing onto their backs on their return to the sea. Additional people gather as the sea is becoming even more active, more alive.

The one minute intervals of humpback breaches are awesome! Two together, then one, then two. Will there be three? How accurate are their biological clocks? How high can this young male, who we decide to name Otto, clear the air water interface? Can we get the first known photograph of such a whale completely airborne? The adults are so large that they only have the strength to rise three quarters of their length from the water. But their resounding re-entries are perhaps the most spectacular natural events in all marine biosciences and ocean observations.

"Ten second warning on the next 'alpha time' for breaching," hollers an inspired Nick.

Then comes the most unbelievable event in perhaps all of behavioural science and whale-human communications. Four adult humpbacks, in their two mixed gender pairs, portray near perfectly synchronous breaches, all four landing simultaneously! Alex falls back on the console seat to be steadied between the legs of George. Nick is making exclamatory sounds in continuous streams!

---

[23] White sided dolphins (Lagenorhynchus acutus)
[24] North Atlantic white beaked dolphins
(Lagenorhynchus albirostris)

Horns blare ashore, more than a kilometer away. The sea slowly recovers from the dynamic whale created waves; then our surrounding ocean returns to its former calm. The air is still, except for the occasional flock of noiseless puffins bolting by.

"6 o'clock, dozens of gannets[25] approaching," announces Nick. These magnificent seabirds, with their two meter wingspan, are called the 'albatross of the northern hemisphere,' and they 'plunge dive' for fish, at high speed, from tens of meters height. Some stall nearby, as if they have seen a school of favourite fish. The 'spinner dolphins' leap vertically skyward, three in synchrony and belly to belly! Between them, trapped, is a small mackerel, and above, astoundingly, comes a diving gannet! Faster than a human can run, the bird grabs the fish and pulls up and out of its plunging dive, just missing the air water interface. The dance enlarges with triads of breaching dolphins and additional hungry gannets, which, swooping downwards for their gifted fish, then beeline towards a nearby Gull Island.

"The dolphins are actually 'giving' the fish to the gannets!" exclaims Otto.

"That's a great observation Otto and we are all witnessing Nature's altruism at its very best," Alex comments to all.

"What is altruism, Alex?" questions the boy, somewhat naively.

"Perhaps the pre-eminent, outstanding word in all languages, most especially in the world of Nature, Otto. Altruism means freely giving without expecting a reward, as the dolphins just gave the

---

[25]   Gannets (Marus brassanus)

mackerel to the gannets. 'True altruism' could mean not only giving, but not even the slightest conscious comprehension of reward. These concepts are foundations of most religious thought on Earth today."

Katrina translates the exact and powerful meanings for the boy, who is scholarly with interest.

"Would you say that altruistic animals, like these dolphins and whales, live closer to a state of 'heaven on earth' than we do?" asks the boy in his native language.

Alex translates for all and then answers, "Yes son, but some humans must surely be included."

It seems hard to leave these friendly animal encounters but indication is at hand that serious feeding has entered the 'NowTIME' of these whales and dolphins. So we wend our way to Bonavista along one of the most scenic shorelines in Canada. A giant cave has an eagles' nest at the entrance and calm turquoise waters with kelp and a pink coral lining its edges. In the very back, air is trapped by the long wavelength swells, and the occasional deep booming sound resonates throughout the cave and into our inner souls. Further along, the signature cries of roosting guillemots[26], fills the silence of the near shore protected waters. A minke swims by on its side so as to possibly count the orange 'Martian-like' exposure suits that keep us so warm and dry.

An island puffin colony lies at the very end of the Cape at Bonavista. We enter the narrows with barely enough room to manoeuvre and with one

---

[26] Guillemots (Cepphus grylle)

engine in reverse we are able to make a sharp right angle turn and emerge out between sheer rock faces, at the very end of the Cape. We turn toward the island home of hundreds of double crested cormorants[27], and can barely view these majestic birds as a typical Newfoundland fog bank rolls in from the east. Turning toward the harbour of Bonavista another minke whale barely altering course, swims by, a boat length away, as it seems to be heading into Bonavista Bay for a 'rendezvous with destiny.' Maybe it will guide us to port! We follow.

"That 'whale contact experience' was a life highlight," exclaims Elliott.

"Compared to some of your other highlights, that's a heart rendering announcement." I comment.

"Can I steer?" asks Otto.

"After we pass through the narrows ahead, Otto. We have an interesting and different day planned for you and Katrina, at Trinity tomorrow."

A peaceful quiet transcends over all and we soon find the safe harbour, safe docking and the friendly, helpful people of this famous fishing port.

---

[27] Cormorants (Phalacrocorax auritus)

Gannets Plunge-Diving

Gannets, the albatross of the North,

From rocky islands they set forth.

Gliding high with their giant wings,

They master two environments, like kings,

By plunging into the sea at breakneck speeds,

They catch their fish with true survival deeds.

With ocean dolphins a fish comes high in air,

A gift to gannets air-diving, which they dare.

CHAPTER 11

Elliston Research

The whale and weather situations look great for pressing theory into practice, so I call William as soon as we depart the *Ceres*. He is anxiously waiting for the call and plans to be on the tarmac at Gander by morning! He has arranged for a helicopter to meet us in Bonavista by 10:00 hrs.

We travel to Trinity having the excitement of possible unfolding events and with detailed talk of deciphering the contrasts between 'whale Time' and 'human time,' if Ida and companions will remain in the 'Elliston restaurant' (as we expect from past experience), and if she will be well enough fed for definitive, low stress 'Rhythm Based CommunicaTions' research. Alex outlines his theories again during the homeward van ride and explains how they now match the British ideas. We are presuming that they will involve, when William arrives, an interesting and exciting reunion of these two great human minds, as well as a potential meeting with perhaps equal but so very different, living, cetacean minds, that are now in nearby waters.

At next dawn, Nicholas, Kirk, Jason and Jay head back to Bonavista on a clear and calm morning, typical of our prime Newfoundland whale research days. They survey the situation at Elliston, which looks ideal, and 'phone home' so that the expedition plans can get underway. After a hearty Dody-Penny breakfast we receive a call that the Gander landing is accomplished, so we can direct a rendevous by 10:00 hours at the helicopter landing site near the small boat floating wharf system in the secluded Bonavista harbour. William arrives and is assisted onto *Ceres* to the 'Admirals Seat,' just forward of the console on the starboard side. Alex and Hans are soon seated forward, with George and Elliot aft, as Nicholas casts away and then jumping aboard stows a hearty lunch, including a few of Chris's special British treats! Another chartered larger vessel with Kirk, Jason and Jay, and with radio contact, will follow behind as backup safety, although the *R. V. Ceres* carries an inflatable, interior life raft, two new independent engines, two anchors, two depth sounders, many life jackets, paddles and flares, first aid and all other required safety and communications equipment. Alex is in charge of explaining the logistics and safety to the slightly nervous but seemingly wonder-struck William.

The near coastal scenery, including spectacular sedimentary geology and lively seabird activity, is a great introduction to one's first impressions of Canada's most easterly province. We round the amazing Cape of Bonavista where presumably Captain John Cabot made a landfall some 500 years before. The lighthouse keeper high above, waves in anticipation of our radio contact. He rather excitedly tells us of a 'heard' of seals just ahead of us on our route to Elliston, with its famous puffin filled 'gull islands.' We pass minke whales near enormous sea caves and giant blows of fin whales further off shore. Now William and all are thoroughly enjoying the magic of the very finest of Canada's east coast marine scenery. As we near Elliston's northern head I explain where the net entrapment of a famous, female humpback whale, appropriately named 'Meg' (after the biological name Megaptera, representing this species giant wings) was located many years ago. Such could well be an introduction to a lunch break story in a sheltered spot, near where Meg was studied by scientists from North America and Europe, before her release. Meg now often swims with Ida's 'heard.' as does her longtime mate Scott.

"Humpback blows ahead, five miles!" announces Nicholas from the mast lookout. "That looks like the same 'South Gull Island Restaurant' where Ida was this morning."

"Great! You are about to go dancing with your first live Great Whales, William. Watch for the humpback's giant white flippers, or as the Greek name would translate, 'giant wings!' Ida's so called arms, are over ten feet in length and she may wave them in the air as a favourite way of signalling, and, for George to photograph! There are Coronula barnacles, which we now believe are sonar producing, on the leading flipper edges. These animals are probably good friends of the humpbacks and were actually first described by Charles Darwin."

We pass tens of thousands of puffins and William is so intrigued that he asks George for a photograph of the awkward water landing antics, and then, most especially their take-offs, sometimes with up to five, 5-10 centimeter long capelin fish in their beaks!

"We are now automatically transmitting our standard 'alpha rhythm' of one short underwater sound pulse precisely at every one minute 'duraTion,' spelt with an upper case 'T' for 'Rhythmic Time,' so you can check with Nicholas if you would like to set the second hand on your watch to this rhythmic 'alpha concept.' Ida's group surely now knows that we are approaching and my guess is that she also knows our range. Everyone watch for rhythmic 'synchronizaTion' which could be any rapid signal, such as an exhalation, exactly when such animal receives our acoustic transmission. Remember that sound travels at about a mile in a second, underwater."

It is time to explain our highly developed, on board communications to William, and also, as a review for the others, the new four pillars of our joint observation abilities. We will need to record, using two hand held computers, all sudden whale movements so that the trusty software can analyze the perhaps hundreds of rhythmic 'duraTions' per hour, both received and transmitted. Such signals as tail and flipper slapping, exhaling, breaching and even sudden turning under the *Ceres* are known, in scientific language, as 'high rise time,' that is fast relative to the 'alpha rhythm' of 60.0 seconds.

"The most important communication now is 'bearing' and we use a standard clock face always oriented with the 12 facing the bow of the *Ceres*. Starboard beam is then 3 o'clock and every hour is precisely 30 degrees so that a sighting at 11:30 is

15 degrees left of the bow line or to our port side. Alex should cover 9 to 12, Hans 12 to 3, George 3 to 6 and Elliott 6 to 9. Nicholas will, at times, cover the stern half from 3 to 9 and I'll take the forward half. The second communication, when you have seen a signal, is the range in metres so that a typical contribution could be: blow, 4 o'clock, 100 meters. If Nicholas or I can estimate a whale's position only, without a visual sighting, we reverse these numbers with a calmer language such as: Ida, 120, 1:30. Notice that after some practice you should be able to not bother with the words o'clock and meters. South Gull Island is now at 2 o'clock, or 60 degrees to starboard, and at a range of 300 meters."

"Down speed from 10 to 5 knots, and notice the slow deceleration so as not to stress the 'heard,' which is a pun on the verb 'to hear' as whales are mainly 'connected' by sound. Such is a suggestion of cetologist Dr. Roger Payne."

William asks, "Are you going to record the vessel speed and direction for a possible chart analysis?"

"Affirmative" says Nicholas, just louder than the soft hum of the amazingly quiet engines.

"Listen carefully for a second signal on the next countdown," as George, watching the console timer counts "5-4-3-2-1-mark." I also announce, "Did anyone notice that we silenced one engine just on the word mark?"

"Double blows at 12 o'clock, range 200 meters, heading our way," says Nicholas. William sits up tall and confirms his very first, live humpback evidence. He later claims this as a highlight well worth his complementary 'price of admission!'

"Blow, 500," says Hans. "Another 11:30, 600," calls out Alex.

"There are now nine humpbacks between 10 and 2," announces Nicholas. "This is indeed 'Ida's family' because I've identified Andrew and Meg from their distinctive tail patterns."

"Down slow to neutral and announce the count downs please Nick." On the next mark the second engine switches off and we are now drifting on this calm sea with the Navy term 'silent ship.'

"'Possible synchronizaTion,' in 'RBT'," I announce! "That's whale #1, swimming with a smaller mate, whale #2. Transmit the 'greeting' Nicholas. This will happen by missing the next 'onTime' mark with 'T' as upper case, or whale time, and we announce two 'offTimes' followed by a final 'onTime,' for our console, acoustic transmissions."

"Elliott, when Nicholas says mark could you drop this pinger overboard and then try to retrieve it, using its cord attached to the boat, within 2-3 seconds"

"A-Ok Peter 'me son'," comes a quickly learned Newfoundland accented reply. There is indeed a warm feeling on board as 'me son' means 'my friend' in 'Newfoundese language.' Elliott is trying to learn all of Canada's languages and he has now mastered at least twenty of the many local Newfoundland words!

"Reciprocal Greeting," excitedly cries Nicholas. "And it's Ida who is heading toward us at 1 o'clock. My computer stress light went from red to orange to green, all in this last six minutes."

"William, notice that humpback Ida is coming directly toward you, now at 50 meters, 40 meters, 30, 20, two whale lengths, turning over, can you see her grapefruit sized eye? Wow! She is now under the boat, and here comes the giant tail. We call that 'perception des personne'!"

For over an hour we gather data with no inclination of human hunger, or for a lunch break. Ida and Andrew confirm Alex and William's thinking of their 'rhythmic informaTion' (spelt with a 'T' for 'Rhythmic Time') encoded in what we have previously called 'whale Time' ('T' as above) now known as 'rhythmic Time,' and quite different from Newton's or Einstein's 'Conventional time.' We have glanced into the 'Depths of Nature,' and we have been humble students, seemingly taught by these 'Great Whales.' These principal teachers soon enter their nearby restaurant, which location now specializes in the cigar sized, post spawning, capelin fish, so that is our cue to head for a protected, cliff side anchorage.

"Up slow to half speed" as I rock the double throttles so as to minimize our acceleration. Soon we are entering the Elliston harbour, and while heading to the northwest side I explain where humpback 'Meg' was scientifically studied by many professionals, during major, nonintrusive experiments in a past July.

"We had a giant maze anchored straight out from this rocky promontory, tents all across that small, grassy field and even took our showers under this waterfall, which still makes attractive sounds for our lunch break background."

In the small caves are pre-spawning capelin taking shelter before they lay their eggs on nearby beaches.

"Most of the 'whale sonar tests' must now be imagined for the winter seas have removed much of the staging that we used for our numerous inflatable vessels and offshore acoustic recording. Heart monitoring was performed from the eastern tip of that rocky promontory, looking somewhat like a large stationary naval ship. Meg was temporarily tethered, but let out by half a mile for her feeding, and taught to swim backwards to the shore for the crucial maze sonar experiments, similar to what had been done with Harvard bats some many decades before. She was cooperative and seemed to enjoy our close contact and touch, but certainly not on her lower jaw where sensitive hairs may have evolved to locate tiny planktonic food on dark nights. Another scientist spent hours in the water watching her flipper movements for his post doctoral research, and we spent days studying her heart rates, the first of such scientific measurements. We photographed her many baleen plates which presumably evolved to filter water out of the mouth during feeding. And a famous Canadian vet, obtained the very first humpback blood sample which later confirmed Meg's shallow diving, compared to her cousins the toothed whales. A French team monitored the offshore sounds while she swam, temporarily blindfolded, through an enormous maze made of telephone poles and underwater nets. Yes she made 'sonar-type' sounds but only when her presumed mate, Scott, swam nearby! Meg in the maze, was never deeper than ten meters and later we discovered that the Coronula barnacles which appear as sonar transmitters, do quite probably not activate at such shallow depths! But negative results are not always negative science as we discovered many other fascinating aspects of this very first temporarily 'domesticized' humpback whale and we were wonderfully rewarded by her return almost every summer, in great health."

"How did you train Meg to swim backwards after her feeding sessions?" asks Alex.

"The senior biologist tugged gently twice on her tether before the first and second multi human 'team tugs' and on the third time no force at all was needed as Meg apparently understood our desires, and swam backwards."

Chris's gourmet lunch was enjoyed with all six of us sitting around the 'picnic basket' in the bow of *Ceres* with Nicholas pouring the hot chocolate and, of course, hot tea for William. A fisherman wandered down from his nearby house and he seemed just the same as when we had lived nearby to him, in order to study Meg, so many years before. He still showed us his altruistic culture by offering a fresh fish for our supper.

"What kind is it?" asked George.

"Fish means cod in the Newfoundland language George. For all other species we use their 'worldly' names, as salmon, mackerel, etc. Walter made almost his entire living from fish, wood and gardening. And he also had a dozen children!"

"Thanks Walter! We've just seen Meg again, over there off the southern island, seemingly healthy and happy! Our regards to both you and your marvelous better half!" Lunch over we prepare for additional 'whale-contact.'

"Anchors away Nicholas, we'll have to show our 'breach button' to all of these colleagues!"

"Just a moment 'doc,' how did you plan to study Meg directly after her release?" asks Hans.

"A very large, whale research, sailing vessel was summoned and a radio receiver attached to its main mast. Our master engineer designed a transmitter to comfortably fit onto Meg's dorsal fin and she swam away, joining Scott and other humpbacks, but at times moving in the very direction that the wind was coming from, and sometimes faster than the tracking vessel. When we next saw Meg she was a normal looking humpback having her own private geographic locations of travel, like all of the other 'Great Whales,' worldwide. At that time the night behaviours of all such animals was almost completely unknown."

"Breach! 12 o'clock about 1000 metres," shouts Nicholas.

"Amazing," says Hans, "maybe they do really have a 'breach button'!"

"Remember that in our concept of low stress whale communication it is not what happens but when it happens, that carries the information. But one has to know the relative 'RhythmicTime,' spelt with an upper case 'T,' and one word, for 'Rhythmic Time,' that is relative to some cyclic 'synchronizaTion' only known to some chosen animals, as a way of encrypting 'informaTion.' And because its this different whale 'Time T' we call it 'rhythmic informaTion,' to contrast from the usual information, encoded in signals, signs and symbols. The relative part may be related to professor Einstein's famous thoughts of 1905 but the exact connection, if there is one, is not, as yet, fully understood. Let's go up speed to half speed."

"What are causes for breaching?" asks Elliott.

"Probably many motives, of which communication and play seem paramount."

We pass South Gull Island with its zillions of seabirds and now fin, minke and humpback whales are all feeding in the same 100 acres of active ocean. The sea surface is calm except for the 'whale waves,' but beneath, a myriad of small capelin are instinctively schooling. And now we can see clearly the extensive, round fish formations on our graphical echo sounder.

"Synchronized breach on Andrew" exclaims Nick. "I'll begin the rhythmic greeting again."

Three minutes later we transmit the final brief acoustic signal and William, who is timing this message exclaims, "Why was that interval three instead of two minutes?"

"That's because we must miss the first 'onTime' window after 'rhythmic synchronizaTion,' and then transmit the '1ate, late, onTime' greeting message during the following two minutes. It took us many trials without figuring that one out, William, until Ida repeatedly taught us this seemingly acceptable way, which now appears universal for many other animal contacts."

"What are possible language translations of the message?" asks George.

"Essentially 'hello'," rapidly answers Nicholas!

After a reciprocal greeting we transmit the 'north?' rhythmic message. Then, both Andrew and Meg join in a double breach in an 'onTime' window immediately after which they swim almost due north towards the famous Bonavista lighthouse.

"What did the double breach mean in this case," asks Alex.

"It meant simply an affirmative reply. Had it been in the 'offTime,' window it would have been a negative response, and at any other 'Time' it would have had no meaning to us but could have been a component in another, independent. rhythmic communication."

"Wait a moment doc," says Hans. "You mean that this rhythmic system supports more than one simultaneous 'communicaTion,' like speaking English and Elliott's French concurrently?"

"Exactly! We now believe that these animals can possibly support five or more simultaneous 'communicaTions,' like English, French, Newfoundland, Cree and Inuktitut, concurrently!"

"And why could you suppose that?" asks our intellectual, charismatic Elliott.

"Because this 'Rhythm Based CommunicaTion' or 'RBC,' theoretical system allows such a conjecture, and because their cerebrum, being far larger than ours, could accommodate the larger processing power, and finally because their social structures seem unrealistically complicated."

We transmit the 'south?' rhythmic message and both whales not only exhale almost together in the next 'onTime' window, but both humpbacks abruptly turn almost directly away from the lighthouse direction. Then east, or pi/2 radians (90 degrees), especially for William; then west 3pi/2, and the whales follow precisely. We then repeat the earlier Elliston experiments for the four different, orthogonal directions, during the next hour, with additional successful results, and William seems dumfounded! Transmitting the 'third year capelin' message, Ida answers with two tail slaps in the 'offTime' window, but following the 'fourth year capelin' message she and Scott exhale, about one half second apart, but adjacent to each other and precisely in the 'onTime' window.

William contributes, "One pair is 'communicaTing' but seemingly all whales are listening!"

"Bingo! That is a strongly suspected attribute of this 'RBC' system. It seems a form of interchanging 'communicaTions' leaderships."

"Would you then suspect other leadership rolls to alter accordingly?" inquires William.

"Yes, mais oui, but according to needs."

We move closer to a nearby beach and Nicholas uses a dip net to catch a few capelin, just as John Cabot supposedly did some 500 years earlier. They are pre-spawning, mature adults, most certainly in their fourth year. Unlike Mr. Cabot we release them back into the sea to pursue their critical mission, of their beach reproductive process. We then move back to the humpback feeding zone, slightly further offshore, and retrieve three capelin, a female and two males that are floating on the surface, probably recently made unconscious from the forceful water extrusion methods of the humpbacks.

"These two are males as you can tell from the breeding ridges along each side, and this third is a female, more colourful, smaller and, as you can see, still able to extrude a few bright orange eggs. She is definitely post spawning and ready to die at the will of her various predators. The males probably died while fertilizing the eggs on the beach, and then drifted out here after a rising tide."

I hold the three fish in the palm of my hand with a male on each side of the female, so that George can photograph them for William, and for all to see the way that the spawning ridges can be used to force the eggs from the females during the symmetric tail motions of their little known breeding process. Nicholas explains that the fresh eggs are often referred to as 'Newfoundland caviar.'

"How do you know that they are four years old?" asks Alex.

That's a pretty good guess out here now, Alex, scientifically based on their size, but now we also have the supporting suggestion from our human-animal 'communicaTions' with Ida. The latter we have done many times before and the former has been laboratory confirmed."

"Wow everyone! There's a mother calf pair at ten o'clock, range 100 meters and heading our way! The calf is behind and on the far side, so watch for the second smaller blow."

The *Ceres*, now with engines off, is 'super silent' in anticipation of important biological photographic opportunities as well as a true visual gift of Nature. These two humpbacks surface and swim around us, apparently curious but possibly notified by conspecifics of our congenial intentions. The calf is a yearling probably born in the tropics about 18 months earlier, and thus capable of cooperative feeding with the mother. Otherwise, if at an earlier age, this calf would probably not be this far north until somewhat later in the feeding season. We very softly send the rhythmic greeting, which is mimicked by the mother humpback.

"Amazing!" declares Hans. "The calf has moved to the inside."

"That's a record for us! The previous fastest time is over ten minutes."

Now a very curious calf comes to the boat, on the starboard side, and is seemingly attracted to William. There is eye-to-eye contact and I would assume some tears of joy. Now with two whales rolling under the *Ceres,* Nicholas is able to determine that this is a definitive mother-son sighting and seemingly an addition, perhaps temporary, to 'Ida's family.' We decide to name the calf Wil, and his mother after our William's mother. Young Wil then seems not only curious, but playful as well, rolling over and over so that George is able to obtain sensational, full frame photographs with a sunny, glistening, and bright background. Ida swims nearby with additional human whale messages, all logged onto our dual computer system. We are indeed again entering into her 'Living Nature.'

Then comes a highlight of highlights! Andrew, with his giant flukes, throws Wil high into the air, tumbling head over tail until a giant splash is so great that we have to shield our cameras behind the console's windbreaker, transparent shield.

"Darn it! I missed that one," howls George.

Then again, and again, each time twenty, thirty metres into the air goes young Wil, apparently in sheer delight. The landing noises, we calculate might be heard by hundreds of humpbacks so that one wonders if this message of apparent play is being shared by many in the larger east coast 'heard.' George gets his cherished picture and so becomes the first photographer that we know to hopefully film a land horizon underneath a completely airborne humpback whale. Alex and Hans beckon to get closer, presumably so that they can get us to take a video of them and the whale Wil, as well as Wil's splashing water landings! We are dealing with two mature adults immersed into their recollected younger years of play! And such has been, and continues to be, not just humans watching whales but also whales, both underwater and in the air, watching us! It is truly 'people watching' as Wil's eyes so often seem to focus on the entire *Ceres*, both these orange suited, rather strange looking sea level characters, and the acrobatic Nicholas, high up on our main, masthead crow's nest. By now we even see others lined up along a nearby shore roadway, some no doubt wondering if we are in danger! However, judging from this and many previous experiences with these marvelous mammals, we are probably safer here than we will be on the drive home to Trinity. The activity dies down and it then feels like the aftermath of a successful Apollo liftoff.

George then surprises everyone by producing a flask of our famous Newfoundland rum called Screech. "Here is for our celebration for the best multi act drama one could ever imagine that Nature could provide for this group of lucky humans." And he shares the rum but turns to me and says quietly, "sorry skipper, you're still our designated driver!"

We are Bonavista bound but not until we successfully send and receive a rhythmic farewell message, and then comes a final breach from Ida as she presumably gazes upon the *Ceres* slowly moving away to the north.

William remarks, "In your work Peter why do you say that this demonstrated rhythmic 'communicaTion' both requires and produces low biological stresses for all concerned?"

"Because 'RBC' requires rhythmic 'synchronizaTion,' and also produces a system whereby it doesn't matter what the animals do but it does matter when they do it. This when however, is a relative rhythmic 'Time,' relative to the a temporarily maintained 'synchronizaTion.' Ida's family apparently knows that our 'communicaTion' rhythm has a one minute duration and this is the first step in such an 'animal-conTact' experience. We can continue this conversation, if you like, on the van ride home to the Village Inn, where a special banquet awaits us all, in celebration of your distinguished visit William."

The *Ceres* speeds off into the setting sun with occasional friendly dolphins riding on our bow wave, as if to escort us into the scenic Bonavista harbour.

CHAPTER 12

Explorer Endeavors

Alex, Hans, George, Elliott and Nicholas are full of questions for William, during an effortless, mostly uninterrupted ride to Trinity, with a soft Tchaikovsky background. William has a "mind of lightening" possibly related to his ecstatic reactions to such an exciting sea adventure.

"Did Dr. Einstein include 'RhythmicTime' in his theories of relativity?" asks Alex.

"Not to my knowledge and neither does it occur, where it should have in the famous Lorentz transformations," replies William. (Please see Appendix II.)

"Wow! Just wow!" says Alex.

"Do you agree, sir, that we cannot travel backwards in time?" asks Nicholas.

"Yes indeed, as then you could cause your parents not to meet and thus prevent your own birth!" answers William.

"But," says Hans, "if one stands on our south pole, the Earth spins clockwise, but from our north pole it appears for communication purposes to spin counterclockwise. Would you not say that one is backwards from the other? And is not the Earth a clock, and temporal concepts simply the mental readings of such a clock? This seems Alex and Peter's cyclical, 'Rhythmic Time' which is now referred to as 'RT' (or 'RBT') and spelt with an upper case 'T,' so as to differentiate it from 'Conventional time' which has, in physics and other sciences, traditionally been symbolized with a lower case t."

"After today's scientific demonstrations, temporal concepts do now appear to be of two types," comments William.

"Then how do we verbally describe the difference?" asks Elliott.

"Well that can be a problem until we modify texts, but for the next few years we will have to use 'Rhythmic Time,' and as well we may need an additional emphasis by spelling it as one word, versus 'Conventional time,' always spelled as two words. Single word 'RhythmicTime' components could be: 'onTime, lateTime, offTime and earlyTime,' along with the important rhythmic concepts of: 'duraTion, communicaTion, synchronizaTion, and informaTion.' It is this last concept that now seems to be a potential foundation of much of low stress Nature, and possibly some forms of organism spirituality," replies William.

"But can the worlds of science and communication easily tell these different temporal types apart?" asks George.

"Perhaps the simplest difference is that 'Conventional time' is a one directional 'space/speed' always counting in a forward direction. It is usually a non cyclical travel time but if it is cyclical then the forward direction of counting is always from 0 to ½ a cycle (a watch's 6) to 1 cycle, to 2 cycles, to 3, 4, 5, and onwards. Whereas 'RhythmicTime' is always cyclical and counts from 0 to ½ cycle, to 1 cycle, which 'Time' becomes instantly 0 again, then onwards to ½ cycle and then to 0 again. And these cycles can be bidirectional as per simply turning a watch back to front or looking at a Ferris wheel from either side," answers William.

"Now 'RhythmicTime,' as used in 'communicaTions,' can have cycles within cycles, as for example, an Earth day can be within an Earth year. Thus the 'RhythmicTime' at which 2 pi radians (one cycle) becomes 0 radians is by mutual agreement between the minds involved. We were using a one minute cycle today with Ida which is a relatively long 'duraTion' compared with our work with foxes and eagles. These 'duraTions' must be such as to minimize the mental thought stresses of the communicating organisms. Thus a critical factor of 'RhythmicTime,' generating 'Rhythm Based CommunicaTion, RBC,' is that one cycle changes to 0 cycles by mutual agreement."

"However, when your mind reads a stationary cyclical clock, there is little spacial difference in the light paths from clock to thought and both of these mentioned dual temporal forms are now considered to be mental thought readings OF the clocks, not mechanical readings AT, ON or IN the clocks. Therefore the 'Mental Vector Processes, MVPs' of both temporal types may travel at the same vector velocities (generally the velocities of light or of sound)."

"Whew!" says Nicholas. "That's new knowledge! Surely then we need different names for the reading AT the clock as it must occur earlier than one's mental reading OF the clock."

"A-OK, Nicholas, and we do have important different names."

"Explain s'il vous plait!" declares Elliott.

"Well colleagues, time in the real outside world, like colour and shape, is a label like the pricetag on your various food items. And such tags have an amount, or magnitude, but no spacial direction or spacial dimension as east, north and up. We call them all scalar labels but unlike colour and shape, time can be a more rapidly changing, or developing, series of scalar labels. How rapidly time labels change depends on whether you are mainly interested in its seconds or, if you were a plant, interested in its days. And because there seem to be two temporal types, there seem to be two types of scalar labels or tags. And so we suggest calling such labels, both of which are innate properties of all clocks, and also follow the concepts of two temporal types: 'timetags of Conventional time' and 'Timetags of RhythmicTime'."

"Explain that again," says William, "and please emphasize your spacial idea. Everyone learns to define conventional time as simply space divided by speed, so if you walk fast you get somewhere sooner than if you walk slowly. Such time certainly involves a changing space."

"Scalar labels, like time, colour and shape are 'moved,' like cargos through 'changing space' from one's outside world or one's objective reality, to one's thought process, but are so transported by moving mass and/or energy bundles or vectors. Once these scalar cargos are moved, they can obtain a directional property, say northeast, and if they then arrive at one's thought processes they are then proposed to be called mental vectors (or 'Mental Vector Processes, MVPs'). Thus the 'changing space' of temporal scalars is produced by 'MVPs,' not by the scalars themselves."

"What do you mean by vector, and then what is a mental vector?" asks Elliott.

"A vector is any quantity, like the velocity of this van, that has both a magnitude, say 100 kilometers per hour, as well as a direction of north, south, east, west, up as for a hill or lifting device, or down, or some combination of these. If the van is moving, like right now, then it is a vector, but once arriving at our Inn and parking, then the van becomes no longer a vector because it is no longer moving in any direction with respect to the Inn, or the Earth for that matter. On the other hand a scalar is any quantity with magnitude only, like colour, shape, mass, temperature, 'Conventional time' and 'RhythmicTime.' Many people say that 'time flows,' but as all temporal concepts are now considered by many to be scalar quantities, they can be transported like a cargo on an aircraft, but of their own ability they cannot move, fly or flow."

Suddenly a bull moose charges at a full 'vector velocity' onto the road and we very quickly carry out two important and rapid manoeuvers Firstly, we pause until the animal seeing us makes his decision to escape contact, and secondly we make our choice and turn so as to hopefully avoid a collision. We almost succeed but the passenger's side rear view mirror, next to William, 'zings' off as the moose's coat cleans much of the dust from the right side of our van. Luckily all is well as this immense but very attractive animal disappears back into the darkening forest. We stop to recover, both the mirror and our serenity!

"That was not a situation for low stress 'Rhythmic CommunicaTion'!" exclaims Nick.

"More like survival of the fittest moose," adds Alex! "That animal was certainly a sudden and speedy thinker."

"Why not some more tranquil Mozart music," exclaims Elliott as he searches our van library. Finding 'Nachtmusik' for string-quartet, he hands it to William who is in the radio operator's front seat. And then again Elliott asks for an explanation of what has been earlier called a 'vector of the mind,' or a 'mental vector.'

"Such an activity, now called a 'Mental Vector Process,' with its acronym 'MVP,' is the combination of a mass and/or energy vector, such as a moving parcel of light or sound, with one or more scalar labels, such as temporary stationary shapes or colours, but also including scalar time labels, like clock positions, always located at the edge of one's 'Event Space' with its acronym 'Essos.' This last concept can be remembered as a brand of 'plural fuel!' And the former as a 'Most Valuable Player,' possibly the 'MVP' for the 'Game of Life,' just as in a common Canadian term used in a game of hockey or lacrosse! 'Essos' stands for one's 'Event Space Sphere Or Spheroid,' a volume in which meaningful events happen, but only those events that are recognized by a 'Mental Thought Process,' which process, by definition, is located near its own 'Essos Centre'."

Alex joins in, "One's 'Essos' pronounced 'Eee-sos' is a volume surrounding one's body, and exterior to this volume is one's future. Non-humans seem to have little or no ability to plan, create or produce concepts outside of this volume, and such human ability to do this is a seeming miracle of our evolution. Inside 'Essos' lies what we call 'NowTIME,' which, along with memory, is what non-humans possess, and so human-animal communication in the new methods of today will probably always be in their present and past knowledge."

"Then precisely what is the 'Rhythm Based CommunicaTion's Present'?" asks Elliott.

"It is the cyclical 'duraTion' of the main rhythm and message being used, which for our case of today was a few minutes. Humans can contemplate a future, but the humpback whales, like Ida and Andrew, have taught us that for such experiments as today, their 'communicaTions' with us remain always within our mutual message 'NowTIME.' Following a certain lateness concept of the first recognized cycle, say half a cycle or half a minute late, commonly called 'offTime,' we must consider it equivalent to any lateness concept of any of the following cycles, all of which are in the whale's 'Now.' And we then suppose that such is probably true for the entirety of a 'Non-human Nature's Now' or of 'Rhythm Based CommunicaTion, RBC'."

"One o'clock everyone, range 300 metres. Do you see a beaver lodge against the far bank of that motionless water?" We slow gradually to a stop mid way along a small roadside pond.

"Beaver at three o'clock," cries out Nicholas. "And moving an alder branch toward the lodge."

"I can see a small tree moving but not the beaver," says George as he lowers his window glass ready for photo action. "I'll need the fastest film as it's almost too dark."

"Two young beavers came out of the lodge. They are probably about to help the adult who is apparently 'bringing home the bacon!' We named these two 'alpha' and 'beta' last month."

"That's the father moving the branch," says Nicholas as he re-focuses his binoculars for Alex to use.

"How can you possibly know that it's a he?" replies Hans sarcastically.

"I can see a distinct notch in his left ear. That's one of our field marks previously studied," replies Nicholas. Then Elliott, our expert on beavers breaks a short silence.

"How can these Newfoundland beavers possibly exist in such a small territory of water?"

"They don't. Behind us is a culvert, which we just drove over. They are known to use this passage to access a larger pond, behind us, on the other side of the road. We have seen the adults dragging branches backwards through the culvert, preparing for their winter food storage."

"Have we time to try 'Rhythmic CommunicaTion'?" asks Elliott.

"We could make time but such human-beaver trials now would be a wasted effort as there is a much higher stress involved with this food procurement activity as so often with their typical 'beaver busyness.' And besides we are nearly home to Trinity where cocktails are waiting and then a celebration banquet."

" Mobile One to Home Base, do you read?"

"Roger, where are you at?" replies Maggie in her Newfoundland accent, while at the front desk of the Inn.

"We are leaving beaver pond number three shortly, see ya, over."

"Gentlemen, the final scenery starting soon, along one quarter of the Trinity harbour shoreline is often described as this province's finest. It was planned in the 1940's that most of the British navy could anchor fore and aft along the main arm and you will soon see far in the distance the harbour entrance lighthouse on our left side."

As we round the turn to the innermost part of this magnificent harbour, we get a possible whale sighting from an anonymous guest! Nicholas is quite prepared to explain that it is the well known 'whale rock' and says a word of thanks, for this indicates an unusually low tide.

"We had an adult fin whale stranding just to our left side now, at the harbour's end but we were able to save the animal and often see it returning in the fall when the herring and mackerel move into Trinity Bay. See the osprey[28] heading back to its nest on the east side. These raptors seem to prefer the fresh water trout in the streams entering the salt water behind us. It's too dark for photos George so you can enjoy all this 'living Nature' for its two eye visual memories. There's the other adult coming in from the opposite direction. There can't be any eggs in their nest, yet. Last year we had a path cut to the cliff top, high above their nest, in order to observe them, and without apparently disturbing their behaviour, but it is very necessary to wear dark green clothing for camouflage."

We turn onto the peninsula in the harbour centre and skirt around 'Trinity Mountain' which is high enough for a good ski run in the winter and which for summer time we call 'whale lookout.'

---

[28] Osprey (Pandion haliaetus)

"In January the humpbacks stop nearby on their southerly migration when young capelin enter the bay, and hanging to ice-clad rock faces we have counted over a hundred just outside the harbour entrance. In a day or two they all leave, possibly together, for the warmer waters of the Silver Bank, north of the Dominican Republic and Puerto Rico. That's a recommended way to solve their need for thermoregulation. There will be a prize or two if anyone can remember the other three principal needs for all living organisms."

"Food," comments George.

After a short semantic debate we agree on: nutrition-homeostasis, health and shelter. And then we arrive with many of the female staff awaiting on the front porch, even with a tray of lobster hors d'oeuvres!

"Cocktails and Newfoundland beer in the bar gentlemen. The eagle has landed."

After another of Dody's amazing feasts we pass around Australian liquor contributed by George from his recent travels, and we play the 'Ocean Contact Highlights Game.' Each in turn, names a highlight (not the highlight) of the Bonavista days, while trying to not repeat anyone's previous stories.

Elliott begins with: "The initial approach when Ida was eyeing William from a few meters range."

George adds: "The highest breach with land beneath the calf as seen through my lens."

Nicholas is next with: "Rescuing the stranded sailors from Flowers Shoal."

Alex who sits at one of the table ends adds: "The five interrogative human-to-whale messages and their five whale-to-human sensible replies."

We wine and dine in the finest traditions of the Village Inn.

Chris who has joined us for the triple chocolate cheesecake, adds, "Nicholas's role in the crew rescue and then the yacht rescue, as well as a safe return of everyone."

And then, William, at table's head, and guest of honour, all the way from the UK for this special day: "The environment in total but especially the clear indication that 'whale Time' is indeed different from what I have studied much of my life, which is now for no better phrase that I can think of, called 'Newton-Einstein time.' I must leave you gentlemen in the morning to head back to Gander by chopper for a noon flight home."

There is a collective emotion of sadness followed by several short, well answered scientific questions for William.

And then Penny asks her type of question, "Mr. William, would you like to bring your family over to our 'whale country' as we have lots of space and lots of entertaining culture?

"Yes indeed, Penny, unfortunately both types of my 'TIME' are of the essence just now.

"Having you folks together with such cooperative lessons by Nature and hopes for new knowledge is a highlight of today, of our sciences and of my life. This information in the form of mental vectors, in what we have called 'Mental Vector Processes, or MVPs,' could also well be the Most Valuable Players in the Game of Life. They must not be confused with 'Mental Thought Processes, or MTPs,' which are of a conscious mind's component that receives the 'MVPs,' and subsequently may cause physical or mental actions related to them. Such mental vectors would seem to also terminate in one's conscious and unconscious minds, where they may form additions to one's memory, understanding, knowledge, culture and instinct. And now with our oceanographic fatigue setting in I must say 'bon soir mes amis' to end the day."

Breaching At Sunrise

1) The sun comes up,
   A humpback jumps.
   Another signals,
   In return

2) 'Breakfast is here,
   With all you want.'
   'Tis 'altruism'
   We quietly learn.

3) But hunger code,
   Is not in jumps,
   Is not in sight,
   Is not in sound.

4) It's in the 'Time'
   The sunrise 'Time,'
   Jumps aren't needed,
   Our study found.

5) And so -- at sunrise,
   A hungry whale,
   Can bang its tail!

CHAPTER 13

Foxie Bear

The next morning breaks with glorious sunshine and students returning from the field having interesting news. The eagles in our 'A-breeding range,' had Alice guarding their enormous nest, while Albert who had been procuring nutrition for their growing chick, was earlier observed to drop, just like a bomb shell, a full sized snowshoe hare, not hitting, but probably surprising, Alice and their youngster Andy. I suppose that one could call the drop a strike, for the nest is about two metres across, with Andy usually sitting in the middle of the innermost side and never yet apparently daring to gaze over the outer side to the 100 metre drop down to the ocean, where the whales are frequently feeding. The student-built blind is yet another 100 metres above the nest at the opposing cliff edge of the u-shaped valley shoreline. All three bald eagles respond to our short communication whistles and light flashes but they probably know little of the computerized human activity behind the blind, a spruce and fir covered barrier.

"Two humpbacks and four minke whales were in the bay between the eagles nest and Green Island," says Justin. "We were using the new 12 second cycle 'duraTion' and both adult eagles returned a reciprocal greeting and Albert, then sitting on a nearby treetop, answered with the 'west' rhythmic sequence when we posed the interrogative message of 'hare-direction?' After that he flew off and we both got hungry for our breakfast, and so returned to Trinity!"

"Well done! Tim will have a look at your computer results and you should both see Dody for some well deserved menu choices. I'll bet that together you'll enjoy a dozen of her blueberry pancakes and the authentic, Ontario maple syrup that Elliott contributed."

Just then Alex, Elliott, George and Hans, in intellectual consultation as they descend the main Inn stairway, enter our diningroom and surround William, our guest of honour, well before his scheduled departure by helicopter to Gander and then by private plane to Heathrow, U.K. Tim, Kirk, Nick, Scott then enter as well as a half a dozen other academically oriented students, especially to listen to William's wise words about our research discoveries.

The first question comes from Alex as William is receiving a coffee refill. "If one's age is measured in 'RhythmicTime, T' which is different from 'Conventional time, t,' then for the famous 'Twin's Paradox,' wouldn't the twin's ages remain the same, even though one twin went off in space at high relative velocity? After all, the Earth should continue to cycle our sun once a year so that the twin's ages would remain identical."

"Alex, yes! This well known 'Twin Story' is no longer a paradox," says William.

Alex continues, "Did Dr. Einstein have any suspicion that there could exist two partially incompatible temporal forms?"

"There is no such evidence of 'RhythmicTime' in relativity theory," says William. "But why do you presume that these two are partially incompatible?"

Alex replies, "One's mind can certainly add 'Conventional time, t' and 'RhythmicTime, T' but it appears, one cannot multiply these two quantities, such as appears in some 20th century relativistic transformations!"

"Example please, my dear Alex!"

"Certainly," comes a rapid reply! Alex then explains that if you drive in Conventional time, or 'space divided by speed,' south from Ontario on a fast, new highway, for twelve hours, then Lake Ontario will rotate in 'RhythmicTime, T' to approximately Lake Baikal, in Siberia. For negligible Earth translation you can add the 12 hours south with the 12 hours east to arrive in 'A China Space,' but multiplying 12 south by 12 east is meaningless. "So what does this do for inertial or constant velocity motion multiplied by non-inertial motion which is accelerated?"

"This is new physics Alex, I'll e-mail you with some thoughts within a week. After witnessing the novelty of 'RhythmicTime, T' when the whales were seemingly in a form of low stress 'communicaTion,' this may indeed influence some relativity transformations," remarks William.

The second question comes from Elliott, as William gently recovers from "China Space!" and

Einstein's theories, while Elliott slowly moves, both to get some more of his favorite homemade granola and possibly to phrase his question.

"How could so many scientists and philosophers not understand the concept of 'Now' when it presently seems so strongly suggestive that: 'All real TIME is NowTIME'?"

"I assume you are referring to the imaginary aspect of future time," replies William.

"Yes indeed," says Elliott. "But also past times, as scalar labels, as for example Ida's breaches yesterday, all of which are seemingly different signals, that are desperately in need of other semantics from events which are happening right now."

"Why would you think that?" asks William.

"Because past time doesn't change. It seems fixed, like one's birth date, whereas present time is active clock time transported to mental thought totally in 'one's Now,' within mind's environment."

"But conventional time has traditionally been thought to have an arrow from future to past," remarks William.

"And such an arrow is that of a mass and/or energy vector crossing the volume from mind's edges to mind's central thought system. Past time scalar labels just keep interpretive changes somewhat like pricetags in an active market and according to the progress of our 'Earth Clock.' Yes, these scalar labels, or 'timetags,' do seem to have an arrow of changing labels but such seems an arrow of 'timetags' and not an arrow of real 'TIME' or of 'NowTIME'."

Then Tim announces our CETA-Research 21st century 'mega concept' as a question to William.

"Sir, what do you think of the concept: 'NO MIND, THUS NO TIME'?"

"Are you inferring that after death a person has no sense of time?" asks William. "That seems rather spiritual."

"Yes sir, but more. When you sleep, daydream or are unconscious, there appears no 'real time mental concept.' Surely you don't think that this plate of Dody's French Toast has a sense of time? And how could there be a time totally exclusive of a clock?"

"All organisms possess clocks," says William.

And now for a semantic suggestion, so large, that, it may transform 20th century physics into 21st century biophysics.

"Yes, but these are known biophysical clocks which cease to exist after death and don't exist at all in inanimate objects, like our piano over here."

Both George and Hans then express the idea that there seem to be no minds throughout all space, especially deep inside the Earth, far above our daylight biosphere, near and within stars like our sun or potentially inside the newly discovered black holes of space! Such places lack clocks and thus they lack real time, as it now seems defined by a mental reading OF such a clock.

"Just because we can't measure time doesn't mean that it doesn't exist," adds William.

"Ah! But time requires a clock, some mechanism to read the clock and most importantly, a mind to interpret such a reading," I exclaim. "Would you agree with: NO MIND, THUS NO TIME?"

"I am more convinced than ever," says William, as helicopter sounds approach.

"Then does it not follow that space and time are independent and that the concept of 'space-time' could be and should be 'time-in-space'?"

"That certainly sounds better than 'space-in-time,' replies William! "I'll write to you all here at CETA about this shortly. I think that you are correct but right now I must bid a fond farewell to you Peter and all of your guests, staff and research students for one of the true highlights of my life and an eye opener for both dancing with and understanding Nature, most especially if there truly exists a different form of communication dependent on a newly discovered different temporal form. The comprehension of Ida and many of her 'heard' was astonishing and I will be looking for students to study biophysics now, instead of physics."

In comes the pilot for his customary relaxed coffee break before his flight to the Gander International Airport. With fond farewells we all escort William to his Trinity departure for U.K.'s Heathrow. Dozens of children have gathered, some for autographs, and William shakes many of their hands guided by the gentle coaxing of their parents. Sounds increase by 100 decibels and then slowly fade into the sky as mention is made by Alex of the success of this fascinating visit.

"Meeting in the main lounge for all research persons, please."

At The Village Inn, research activity is regained by 10:00 hrs. and vigorous data analysis is happening in the large bay window connected to the main meeting lounge. Tim explains the software printout to Alex and Elliott. George is showing his finest photographs to all and Hans is returning from his daily dash to the summit of our 'Trinity Mountain,' almost big enough for a ski lift in the winter!

"We'll have one or more days of activities here, while Nicholas trails the boat back today and then tomorrow moves it northward so as to intersect Ida's 'heard' as they migrate up the coast toward Labrador."

Hans enters, slightly short of breath, and announces that there are many fin whales in the harbour mouth as indeed Chris had just previously learned from a friendly 'fisherperson's' phone call. Assignments are made for our remaining research vessels and so also are the beaver, eagle and fox programs organized.

George and two students head quickly for a smaller zodiac which Peter will try to navigate for the difficult photo identifications of the subtle crescent shaped patterns on the fin whale's dorsal surfaces. It will be highly advantageous to place the zodiac up sun and within a few metres of these 20-30 metre, second largest of all living animals. They are friendly, inquisitive and not feeding and we do get ideal identification photographs so that George is again ecstatic with his potential contribution to the North Atlantic fin whale research of population dynamics. The students are impressed by the high narrow exhalations that we so often describe as looking like a giant, royal palm tree. Inside, the narrow harbour entrance, which is only a few whale lengths across, a naive student comments that with a suitable net we could have ourselves a captive fin whale family!

"And where 'my son,' meaning my friend in Newfoundland language, would we obtain the dozens of tons of food to keep them in our preferred low biological stress?" Then the fin whales depart the harbour and we return for planned fox communication research.

"Fox research is best done around noon on a warm day when the kits usually come out with our 'Rhythmic CommunicaTion' invitation while the adults are off hunting for family feeding."

This planned expedition is to a backwoods logging road where Joanna had previously found a sizeable den excavated partially under the road and alongside a scenic trout stream. Once again Tim will head up the computer programed communications data while George prepares tripods for long range camera equipment. Dody, a critical member of the pre-departure team has ample lunch prepared with even a few oatmeal-raisin cookies as fox treats! Our rule is to keep fox rewards less than two percent of daily nutritional needs. This amounts to no more than 1/4 cookie, per fox kit. Others join with the light-hearted assignment of inventing suitable names for the kits, preferably ones for future recognition.

We head out in late morning in two, four wheel drive vehicles and soon enter the logging road which passes a large beaver lodge where the occupants are expected to be fast asleep until the early evening. We stop 100 metres before the den for a customary pause to check for signs of the adult foxes.

Joanne and Tim go quietly ahead until an excited radio voice announces, "Dad, there is a black bear trout fishing in the brook, not 20 metres up stream from the den!"

"Great, can you start the alpha signals at 32 seconds so that your late durations are eight seconds apart?"

"Roger," replies Tim.

George is now focused in from 50 metres and we can hear multiple photographic shots every time the bear catches and ingests a trout.

"Why don't we call him 'Foxie Bear' for his association with the den," exclaims Alex.

"Or just 'Fob' for short," suggests Elliott!

The first curious kit emerges, grabs Joanna's hand held computer and then starts back towards the den entrance. "Good Lord," exclaims Joanna as she lunges after the fox causing him to drop the computer just short of his horrifying hiding hole! "Did you see the black tip on his tail Tim?"

"Yes indeed, lets call him computer kit, if you like" replies Tim.

"Tim to all. We have succeeded in getting the reciprocal greeting returned from the bear so we can go closer and so can you."

"We're calling this guy 'Fob,' for 'Foxie Bear' exclaims Alex and does it seem like 'Fob' has finished his fishing?"

"Yes, and he's crossing the brook, hopefully for a low stress contact," replies Tim

Tim sits on a large log. Joanna is not far away at the largest den entrance. Everyone else is crouched beside and behind the closest of the two vehicles. 'Fob' as now expected, because of the low stress of 'Rhythm Based CommunicaTion,' comes over and sits near to Tim. The 'communicaTion' heightens with head and arm motions by the bear and hand and arm motions by Tim.

"What are you learning?" asks Elliott.

"This animal is from Bonaventure Head, and is last year's cub and, likes Dody's cookies!" replies

Tim. "See if George can photograph the left leg white patch and even the handsome bluish eyes."

Next, Peter and Elliott slowly move closer until both are able to read the clock times on Tim's computer. The bear stress is exceptionally low and this can be correlated with the seeming satisfaction of both 'Fob' and Tim while using the truly amazing 'RBC.' We are indeed producing low stress as each participant is completely free to use both high rise-time and low stress signals, such as a human finger snap or a short clicking noise made with the tongue.

Meanwhile Joanna has befriended two kits and has already verified 'synchronizaTion' at a 20 second rhythmic interval. George has moved so as to get the foxes and the bear in the same frame.

The fox kits, now seven in number, become extremely playful, partly because it is warm, but also because they all seem to enjoy the exercise. Such behaviour is found only in young fox kits. Except for Tim and Elliott, we all lounge about the main den entrance. We try to name them based on appearance and behavior. "How about 'Alpha, Beta, Gamma,' etc., in order of size, but we need to note the three partially black tails as well as this 'Alpha' character's desire to crawl across Joanna's shoulders."

Then 'Fob' wanders over and cuddles the 'Alpha' kit between its arms!

"Wow! photograph that one George," says Alex.

Then, far ahead on the logging roadway, and carrying a snowshoe hare, there appears an adult red fox, apparently an early returning parent. 'Fob' is relieved of his pet and we all move carefully away. When the kits see the adult they scurry back inside the den and we all retreat to the cars to watch. The adult is large and beautiful with an enormous red bushy tail. Into the den with this substantial meal rapidly goes this big bushy tail as all humans and the black bear watch with interest. The communications rhythmic research is over as the kits hunger stress has heightened and there is no time for rhythms when it's meal time.

Both cars are forced to drive backwards for what seemed an eternity, before a turn around place appeared. Then onwards to the harbour lighthouse for a picnic lunch, while hopefully watching the fin whales.

We pass the old whaling station in Trinity harbour, from the early 20th century, and then rising up over the last hill, suddenly there appears miles of the Atlantic Ocean, but no movements of marine mammals.

This setting, however, gave us an ideal site in order to enjoy both Dody's lunch and at the same time a warm, sunny, relaxing ocean view.

"What did the fox kits tell you Joanna?" asks Alex.

"Interestingly, all but one switched to the low stress of 'RBC' rather quickly and then three answered the rhythmic greeting with different signals. The first used ear and tail twitches, the second used foot stomps and the third jumped onto my lap and dug his nose into my bare arm. Perhaps he acquired this friendliness from watching his two siblings," replies Joanna.

"Or perhaps it was because of your calm friendly and experienced demeanor," says Elliot.

"Perhaps, but definitely combined with the 'Rhythm Based CommunicaTion,' answers Joanna.

"Look everyone. Dozens of harp seals and all apparently looking at us!"

"They remind me of small dogs trying to get their heads high above the water," comments Hans.

"Watch carefully now, by command from a leader they should all dive simultaneously, as in right now. - Gone!"

"Now there's a flock of sandpipers[29] on the beach, like the previous turnstones. There seems no way that they could navigate as a coherent flock without 'Rhythm Based CommunicaTion'."

"How does that work again?" asks Mark. "Can you give specific numbers to make it more realistic."

"With the whole flock quickly establishing rhythmic 'synchronizaTion,' as of wing motions, the leader is say 20 milliseconds late on a down motion. Let us say that down could mean turn left and 20 milliseconds could mean 20 degrees. The remaining command is an upswing whereby everyone moves to the right together. Such is a combination of a signal, 'SBC' of down and a 'Time, RBC' of 20 milliseconds. Otherwise there is no time for purely 'Signal Based, SBC,' orders."

"This is an example of a scalar label of 'RhythmicTime, T' combined with a scalar angle, with magnitude only and with displacement components. The distance of travel is determined separately by the leader," says Mark.

"Excellent! Temporal concepts are scalar labels with magnitude only, and as such, they can only

---

[29]   Sandpiper (Artitis macularia)

be combined with incoming 'mass/energy' vectors within or outside of any mind."

"What are all the small boats doing in the harbour mouth," asks Hans.

"Would you folks like fresh cod for dinner, this is the opening of a near shore fishery?"

"Yes indeed and with your barbeque technique!" replies Elliott.

"Ok! Lets go for it. We can leave our dock in about an hour. We should be able to catch five fish in possibly ten minutes."

"You must be kidding," replies Alex.

We leave the dock at 4 p.m. with five fishermen, tackle and bait and in a few moments we have landed a five pound cod, the first of many.

CHAPTER 14

Heading North

Nicholas and Kirk have left early, trailing our largest inflatable with its lockers chock full of supplies and equipment. Tim, Scott and Justin leave next for northern destinations for caribou and bear research. Hans, Alex, George, Elliott and Peter are preparing to leave by helicopter to track Ida's group, firstly off the Greenspond peninsula, then around Fogo Island and then on to Twillingate. We have excellent radio contact with the two road vehicles and lots of logistical and radio support in the friendly outports of northeast Newfoundland.

We search the Greenspond area, then pass Wesleyville, Lumsden and Musgrave Harbour with only a sighting of ocean dolphins apparently riding the bow waves of two fin whales. Next, on to Fogo and Change Islands with no humpback whales but lots of giant icebergs.

Then out of the starboard window Alex sees a live polar bear wandering on a large iceberg! We each use binoculars for this sighting as the pilot circles around the enormous bear. I then place a routine call to Twillingate for the availability of a polar bear rescue team.

"Calling Jabez Freeman; Peter here. We are in a 'chopper' 10 and 1/2 miles northeast of you searching for Ida's group of humpback whales and we have just sighted an adult polar bear riding the largest berg in the area. Can you send out a bear rescue team, and did our humpbacks show up in your area?"

"Hello Peter. Yes we have the right vet available for a bear rescue and we now have lots of humpbacks and capelin just off our lighthouse," replies Jabez.

"Thanks, we'll land near the launching dock shortly."

"Come in Nick. Did you read that?"

"Roger Dad, we can be there within the hour."

We land at Twillingate and as Nick has predicted, within just over an hour, we are headed out of the harbour to hopefully find Ida. The lighthouse is on an enormous cliff top and under this are both schools of the 10-12 cm (4-5 inch) capelin and many feeding humpback and other whales. We didn't find Ida right away, she found us! Our rhythmic greeting signal spread throughout the group (also called a 'heard' because they are mainly connected by sound) and both Ida and an old favourite accompanying male, named Andrew, came right to the boat, both apparently having been well fed. The reciprocal rhythmic greeting followed and Ida finished with a very close half breach so that George claimed that he may have filmed the barnacles, named and accurately drawn by Charles Darwin. They were seen aligned along the forward edges of both of Ida's long flippers.

"What is the biological purpose of these barnacles?" asks Hans.

"We've studied them for years both on and off humpback whales and they certainly seem like sonar transmission devices."

"Then what triggers the sounds? Is it depth or temperature?" asks Alex.

"Neither! We believe that it is 'Rhythm Based CommunicaTion, RBC'," I reply.

"And what do the barnacles receive in return?" asks Elliott.

"They get taken to the best restaurants in the North Atlantic Ocean, and without predators." I explain.

We gather data from Ida, firstly about the other new whales in the 'heard,' and then about what she might know of their food stocks both here and further north.

Minke and fin whales join a feeding frenzie and later we are able to explain the considerable species differences to some of the cliff watchers, including a relatively new light house keeper.

Following this field research we celebrate in my friend's restaurant after dusk when the humpbacks have headed northwest toward the Horse Islands. That is an indication that a main whale restaurant is quite possibly in White Bay, so that Nick and Kirk prepared the boat for an early morning departure by road, to Jackson's Arm.

We give Nick a half day head start while we spend several hours in the famous craft shop and museum at Twillingate. Then by 'chopper' we follow the whales, rounding the famous Fleur de Lys and spreading out in search for capelin, their preferred form of nutrition. Half way across White

Bay we spot both Ida's group and capelin schools, which means that Jackson's Arm would be an ideal location to launch the *Ceres*.

Then by radio, "Peter to Nick. We'll arrive as planned about mid afternoon. We'll need accommodation for five, plus yourselves and hopefully we can head out as early as convenient in the morning."

"Roger Dad, we'll be dockside in about half an hour and I'll give Stan Petley a call for his logistics recommendations. I recall that he has family there."

We landed with Patricia Petley and her sister Susan welcoming us at the wharf in a customary Newfoundland, outport friendly style.

"We have room for all for the night and Stan is here as well," says Patricia.

"Wonderful! Let us take you all to dinner. I see a lot of lobsters being unloaded, just there."

"Ok" says Susan, "We'll supply the cooking if you bring along as many lobsters as you fancy."

"And is there any place in town, at this hour, to get a bottle of 'Screech,' such a great local rum, especially for our mainland guests and everyone?"

"Sure, just up from the wharf on the north side," replies Patricia.

"Great! We may need a few for the possible long trip up to Labrador, where the whales seem to be headed."

Patricia, her sister and their father Stan, and families poured on a home style feast with exciting stories about Stan's secluded retreat just up the coast, with its moose, salmon, black bears and isolated, wild 'Nature.'

Alex asked Stan if this 'Nature' was always friendly, and so came the story of the hungry bear that tried to break into their food storage locker, but failed when their Labrador Retriever woke with the noise, and barked the bear back to its hillside resting place. How a 200 kg bear could be afraid of a 50 kg dog must have been related to a 'dog-noise' factor!

The wind died so that we could hear the whales exhaling, just a few miles offshore. Two fishermen with capelin traps nearby had expressed concern of whale-net collisions, so we gave Stan two sonic devices from onboard the *Ceres*, that had been tested to keep the whales back without changing the fish behaviour. Instructions mentioned were to attach them to the trap 'leaders' about 10 metres from the door where fish entered, and to remember to disconnect the batteries when hauling the trap at seasons end.

We all slept well, especially our foreign guests who were again amazed at the typical outport hospitality.

The next morning we invited Stan to join us for the intended acceptance of his invitation to stay at 'Little Sandy Harbour,' he and his wife Maria's dream of 'Nature's Wilderness.' Nick then loaded the *Ceres* with fresh food and the rum.

We studied 'RBC' with Ida and her 'heard,' with fin and minke whales, with ocean dolphins, and we even tried our hand unsuccessfully, with small groups of seals. When we met one of the trap fishermen he was thankful for the protection device and promised to mail us his catches before and after the pinger's use.

Finally we arrived in 'heaven' with its chosen name of 'Little Sandy Harbour,' and with Maria Petley waiting on the wharf. Osprey's swooped into a shallow section of the harbour, often emerging with capelin, probably planning to feed their nearby growing chicks. Eagles glided high above, in seeming pure enjoyment while salmon leapt up the rapids at the head of the harbour, Maria had caught two large salmon which were prepared for dinner using her favourite marinating and cooking recipes.

Another special hour was approaching in which to open a second bottle of rum but before this there was a challenge, Nick's team versus Kirk's team, for a game of beach volleyball. Stan and Maria acted as judges but Alex and Hans had obviously much previous experience and extra height. Elliott became very busy digging for clams, one of his favourites, especially with fresh salmon!

Story after story followed an amazing meal, which included Maria's garden fixings and very experienced berry pie baking. The story receiving the most attention, questions and interest was Stan's so called 'moose invasion' of the previous summer. Down the slope from the western highlands came all together, seventeen moose, including several calves and one could not imagine how they could have made the descent. They appeared friendly except that Stan was needed as a 'scarecrow' to protect Maria's garden. They very nicely cut the overgrown grass beside the cottage, drank basins of water placed outside for them and allowed friendly grooming of the calves! With no more obvious food available they ambled back to their home and left the Petleys and their guests in curious amazement.

Early the following morning we followed the whales northbound, waving a fond farewell to Maria and Stan, but carrying a fine invitation to stop again, if possible, on a southbound trip back to Jackson's Arm.

Past Greater Harbour Deep we went with a lunch stop at the ancient whaling station of Williamsport. Some of our crew collected a few smaller sheets of baleen which had been weathered clean and dry for decades. Past Englee and Conche we traveled to the amazing Fishot Islands. We went ashore to find flat meadows filled with abundant wildlife, and surrounded by an open ocean environment on all sides.

Then we crossed Hare Bay to perhaps the largest cave in eastern Canada. We entered, being wary of the changing heights. Further and further we headed westward and into the cave until in the distance there was a faint cry for help! We passed a sunken trap skiff and there beyond was a living, bedraggled human crying out for rescue.

"Sure we'll take you home, what is your name?"

"Sean Cook from St. Anthony," he replies. "My mast hit the rooftop three days ago and there is my boat, sunk to the bottom."

"Keep an eye on our mast Nick and tilt it back to the engines if necessary."

"How about a spot of Screech Sean followed by some food."

"You fellows saved my life," comments Sean.

"You look pretty roughed up my son, I think that we ought to get you to the hospital in St. Anthony as soon as possible."

Sean sits quietly in the stern, wrapped in a blanket and seemingly somewhat relieved emotionally.

We arrive shortly at a dockside ambulance in St. Anthony and because of our radio warning, Stan's wife and four children are waiting with outstretched arms.

Soon a friendly fisherman inquires of our mission, and we described that we had missed the whales which he had seen on the outside of the Grey islands, not inside where we had travelled. So we head back to find Ida, Andrew and the whole group feeding east of Groais Island and welcoming our reunion with energetic behaviour. We return to St. Anthony for the night but before long we were immersed in a celebration party for Sean's safe return.

"How did this accident happen?" asked a local doctor.

"Sean claims that he drove onto the beach for a lunch and when he fell asleep, the tide came up and his mast put a hole in his boat. Then followed three lonely days and three cold nights on the beach several hundred metres from the seemingly small cave opening. Search parties had failed to consider that he could be surviving way back in this gigantic cave. His wife had never given up hope, however, especially as no sign of his boat had surfaced. She had persuaded the search crews to comb every island, looking for flags or other signs."

From then onwards we regarded *Ceres* as the ideal cave exploring and rescuing vessel, with a draft of a few inches, once the engines were lifted and with the mast capable of being lowered almost to the horizonal.

Needless to say, once again we needed no paid accommodations and we even acquired a guide for the next morning for the trip round the northern Cape Bald, and hopefully onward to more whale restaurants and to the world heritage site at Red Bay Labrador.

We met Ida's group also heading north seemingly to Labrador but this time we would try something unique, a communications first.

Just north of Cape Bald on Quirpon Island, the most northerly tip of Newfoundland, Ida and Andrew approached the *Ceres* in a rather more curious manner than we had previously encountered. Well into our most successful 'Rhythm Based CommunicaTion, RBC' we asked where they were headed and received a puzzling response of uncertainty. We then notified them of our intent to go north and westward to the south Labrador shore looking for the biodiversity of upwelling, although we did not have the rhythmic vocabulary available to explain the physical oceanography to them. They accepted the idea of following, possibly based on a our ability to find capelin and other food with our different sonar techniques, and, perhaps their millions of years of discovering the Labrador shore upwelling by trial and error or connected possibly with their education.

We kept our speed constant while crossing the Straight of Belle Isle making sure that Ida's 'heard' was able to follow behind.

On nearing West Point Labrador we struck massive upwelling including plankton and fish, the very biodiversity that had attracted the whales here for Europeans to notice from early in the 16th century. In fact one Portuguese log book jests at humans being able to walk across the Straight of Belle Isle, from Labrador to Newfoundland, on the backs of the whales!

We leave Ida's 'heard' off Saddle Island as we enter the now famous Red Bay Harbour.

CHAPTER 15

Red Bay

Red Bay Labrador, perhaps the greatest archaeological discovery of the 20th century, awaits our investigation and concept of its importance in the early European exploration of North America. From archaeological evidence it appears that fully rigged 'ships of the line' plied the North Atlantic less than thirty years after Gasper Corte Real discovered the magnificent harbour anchorage in Trinity, on Trinity Bay, Newfoundland. This event was less than 10 years after Columbus gathered so much credit for claiming his 'discovery' of North America! It seems no accident, that Gasper Corte Real crossed Trinity Bay and then found the safest harbour imaginable, for such crossings of the Atlantic desperately needed such a harbour, and rest. The safe harbour at Trinity is based on nearby land on all sides, protection from ocean waves, which do not normally make right angle turns, a reasonable depth, and access to shore with either no natives or friendly, perhaps trading natives. It also seems no accident that Gasper Corte Real made his father's potential landfall at Baccalieu

Island, and recognized it from the thousands of gannets, enormous white, black and yellow seabirds, plunge-diving into the ocean to capture their prey at perhaps ten metres or more below the surface. There are few other gannet colonies in the North Atlantic and none that fit the fact that Gasper appeared to have sailed directly across Trinity Bay, presumably at his father's suggestion, into Trinity harbour. From only a few kilometers outside of this harbour entrance, an explorer with no previous knowledge would not be able to recognize the magnificence of such a safe harbour lying just ahead. And Gasper's father, Jose, was known from Portuguese archives to have landed and returned from such places in 1472, twenty years before Columbus! Trinity, on Trinity Bay, what a great place to get the much needed rest when arriving from a treacherous crossing of the North Atlantic.

Trinity, was also a great place to come for rest when planning to return across the Atlantic in the 1500's, in order to bring back enough whale oil to light the lanterns of Lisbon, Madrid and Paris, from the developing industry at Red Bay, Labrador. By about 1550 Red Bay appeared to have become the largest business in North America, for quite possibly centuries to come. The upwelling, caused by the Earth's rotation (Coriolis force), affecting the eastward flow of the St. Lawrence River may have created one of the largest and most successful 'world whale restaurants.' This upwelling of river waters brings up the phosphates and nitrates from the sea floor, allowing proliferation of tiny plankton and the larger organisms upon which 'The Great Whales' feed. Even today such mammals are seen traversing the north shore of the Gulf where they can essentially get their "one meal, lasting for many months, per year," before their southerly migration in advance of winter ice and the disruption of this amazing biology.

Red Bay is a small but protected harbour where the harpooned whales could be deposited next to Saddle Island. Here they would be processed using many discovered metal caldrons, and down from the summit of this island would be delivered the finished barrels for the oil shipments to Europe. So the barrel maker had an important role to play and in the early years, he (there were never any females allowed at this time) claimed that in order to procure a stool, he had found an immense double vertebra to make a comfortable seat! Now such bones are rare and represent a very old whale, perhaps one of the first one's killed because of its size and/or slow swimming. The recent recovery of such a double vertebra verified the presence of a truly enormous bowhead whale[30], perhaps 400 years old!

We tied up *Ceres* at Saddle Island, offloaded and went straight to the summit carrying several sets of binoculars. There in the foreground was Ida and Meg, with Andrew, Scott and the remainder of what we have called Ida's 'heard' (or group), not far beyond. Capelin flew out of the water, followed by spectacular "lunge-feeding," where the whales emerging with open mouths, prepared to filter out tons of water through their baleen plates, while ingesting quantities of their favourite food fish.

"What a perfect place to observe this astonishing behaviour," comments Elliott.

---

[30] Bowhead whale, (Balaena mysticetus)

"And, right where the barrel maker worked for much of the 16ᵗʰ century," adds Nicholas.

"Then what caused the termination?" asked Hans.

"Three words ended this Red Bay story, --- The Spanish Armada!," I say.

"Do you mean to say that all of that lantern fuel was cut off by botched war plans?" replies Alex.

"Yes, and as far as we know, no more whales were sacrificed in the Red Bay operation. Come on down somewhat and I can show you a grave yard, probably used for more than a half century. This industry seems not to have been covered by life insurance!"

We survey a small grave being actively excavated, because clothing and artifacts can be of historical significance.

"This was a lad, not even full grown," says the researcher.

"How can you be sure that it wasn't a young girl?' asks George.

"Simple, the European literature repeatedly prohibited females from Red Bay!"

We descended further on the harbour side of the island in order to get a view and description of a major shipwreck location from which many artifacts had been removed. This vessel, probably left behind over the winter, and then damaged by ice and wind, lies beautifully preserved in these colder waters of Red Bay harbour. Associated also are some of the smaller whaling boats, one of which has been skillfully reconstructed.

We visited the museum and at the same time arranged for nearby meals and accommodations. The item that excited Elliott and Hans the most was a reconstructed wine glass said to have been found in pieces within the captain's cabin. We wondered what might have been the circumstances of the divers devouring any of the palatable wine! Searching through what could have been an acre of bones and artifacts, we found mainly biological evidence of bowhead and right whales[31], which being the slowest and perhaps the richest in oil, would have formed the foundation of this historic industry.

After a late lunch, Nick, Alex, George and Peter went out for "Ida-RBC" research and received what perhaps could be interpreted as a thank you for bringing the whole 'heard' to this traditional and first class whale restaurant. Ida came over and rested her head on the side of the *Ceres* but we had to warn Alex and George not to touch the tiny hairs on the anterior part of her lower jaw. Otherwise we could stroke her gently, which apparently seemed to be enjoyed!

"Why not the lower jaw?" asked George.

"Because these structures are not ears, as on the upper jaw. They seem to be delicate sensors to warn the whale of increases in biodiversity. You can, however, tickle her baleen plates."

The next morning we arose early, said farewell to Ida and her friends and headed south, hoping to meet the next whale group coming up the Great Northern Peninsula. Off the Gray Islands we met and researched three species actively feeding, but because of the weather and involved logistics, our journey then took us rapidly back to Little Sandy Harbour, to Maria and Stan.

---

[31]   Right whale, (Eubalaena glacialis)

"Here is a vertebra from Red Bay, Stan."

"Tim, Scott and Justin are camping on a meadow, just over that ridge, doing your communications research with our black bear families. They were told not to take any open foods and I suppose that they may be getting hungry!" says Stan.

"Wow! Thanks. Nick and I will take them up some supplies and check their data."

We depart with loaded backpacks while there are signs of a beach volleyball game getting underway.

"Tim from Peter. Do you read? Nick and I are on our way up to your place with extra supplies."

"Hi Dad! We're getting great recorded data from not one but two male yearling bears and there are also two adults seemingly involved at times."

"We packed some food but nothing with an attractive scent. See you soon."

"Great! - Bear left (no pun intended) at the hilltop. We are on the far side of a two acre meadow with a large green tent plus a smaller beige one."

"A-OK! One bears left while looking for two bears right!

"Ha! You certainly sound like my Dad!"

We soon saw the meadow, the tents and one of the young bears playing with an old tire! Then the second yearling appeared from behind the tent and joined in this seemingly healthy sort of play. Nick and I sat in wonder at the forest edge, trying to contemplate what sort of rhythmic vocabulary had been developed and if it involved any population dynamics. Soon the bear placed the tire near to the tent and both bears, seeing us, moved back to the meadow's edge in anticipation of Tim's message that we were friendly. They then saw our father-to-son greeting after which they mimicked this by a bear-to-bear hug while Tim kept up the alpha rhythm and the 'RBC.' Scott appeared from the smaller tent, where he had been watching. He grabbed the tire and rolled it along the meadow near the two young black bears. Both cubs advanced and continued their play with the tire while an adult remained in the background obviously approving of the youthful antics. This was not play-fighting or kicking but the much more mild form often found in siblings and called object play. Such is now known to be very beneficial for the development of both minds and bodies. It teaches both flexibility and coordination. The friendliness and low stress developed by Tim's 'RBC' work meant that it seemed safe for Scott to grab the tire and then roll it far enough for each cub to have a challenging race to reach it before the other!

"How many adults are there?" Nick asks..

"We have counted only two, one of which is watching us from that clearing over there," replies Justin.

"How did you get here?" asks Nick.

"We got a fishing boat ride up from Jackson's Arm, at the suggestion of Stan's family, yesterday, after a great session with a caribou herd north of Deer Lake. I have a photograph of both Justin and Scott, each next to a young caribou. The adults stayed close and interested."

We helped Tim and students to finish up and then to bring their gear down to the *Ceres*, except for the sleeping bags which would end up in the grand recreational room after dark.

We recounted stories of our adventures during another fine evening with Maria and Stan enjoying the peaceful and natural scenery including much shore bird activity.

The next morning, amidst feature farewells, Stan came to the dock rapidly with a fisherman's news that the southern portion of White Bay, near Jackson's Arm, is unbelievably crowded with herring[32]. We sincerely thanked Stan, and Maria, before heading south in hopes of finding some last cetacean treasures, like fin whales who prefer herring. And we were not disappointed as five miles away there were six enormous palm tree shaped fin whale blows seemingly forming their corral, just as they did with the smaller capelin fish, back in Trinity Bay.

"Look behind," shouted Justin, "dozens if black dots like pilot whales."

"I doubt they are pilot whales which specialize in squid and are here later in the season - But - hold on - these are twenty or thirty 'orcas.' Note the enormous dorsal fins on the older males. Soon they might surround the *Ceres,* so keep all of your arms on board until we investigate the behaviour."

"Another similar Orca 'heard' is approaching from the east," says Alex.

"Hey! The fins 'G.L.' and 'Scratch' are here, arriving from Trinity Bay," shouts Nick.

The orcas[33] pass us by with the young ones coming closest, perhaps to satisfy their curiosity. The older ones swim to observe the fin corral and then after ten minutes of random chasing they decide to copy the fin whales.

---

[32] Herring (Clupea harengus)
[33] Orca whale, was killer whale (Orcinus orca)

We approach a spectacle never yet reported, with about forty adult orcas circling the bait and about ten smaller ones on the inside going in the opposing direction. George films away trying to stretch what remaining film that he has for this unsuspected spectacle.

The orcas used an additional weapon in that every few moments an adult would leap in the air with a deafening splash. At that time we noticed a partial changeover between the corralling outside whales and the interior feeding ones. Then 'G.L.' and 'Scratch' approached to observe this unusual spectacle.

After feeding, the ocean became friendly with curious gestures of inspection of our vessel and manpower. 'Scratch' even swam past by a few metres to windward but seemed careful to blow his many litres of water just ahead of *Ceres* and George's cameras.

It seemed as if we were 'Dancing With Whales' right into nearby Jackson's Arm with the ten of us being emotionally drained. We landed and hauled the *Ceres* for trailing and then with Nick driving Tim, Justin, and Scott and trailing the boat, the remainder came behind in the other vehicle as we headed for Trinity.

Nearing our home port, we called ahead with the news that all is well and could someone mix up the famous Manhattan cocktails, complete with all the secret ingredients which Chris would surely know. I announced that there would be a prize if anyone could name two of these secret ingredients with the only hint that one of them came from the same port from which John Cabot sailed to Bonavista in

1497. We arrived together to a large and very warm welcome.

Joanne and students had completed extensive 'Rhythm Based Communication, RBC' (or 'CommunicaTion') experiments with our local beaver family, with three nesting eagle families, and with two fox dens, one familiar near the kittiwake colony. We were anxious to hear of the results. We traded stories in the bar before, during and after a celebration lobster banquet for all, complete with fine wine for those of age, and much laughter for all.

"The beavers, named Charlie, Coleen and their new offspring Cory and Coby were fast to utilize a large rhythmic vocabulary from last year, and the young beavers even initiated some of it during our second encounter," contributes Joanne.

"We received the 'reciprocal greeting' from Cory on our second meeting, upon his emergence from the lodge, which means just following his long daily sleep. This 'RBC' required early 'synchronizaTion,' which was sensational, and produced low stress, which was spectacular. We had numerous hand contacts with all four but mainly with Coby."

Tim jumps in to describe his bear researches including the extensive play behaviour. These antics dissolved at first when an adult approached from the woods, although with additional 'RBC' the production of low stress and play resumed.

Alex then described Ida's placing her head on *Ceres* and his studying the tactile structure and baleen with hands-on biology!

"Two of the eagles landed by our blinds and one seemed to want to give us a squid[34]," adds Mark.

"Wow! That seems early for the squid northerly migration in the Gulf Stream, and fresh squid is our most popular fall menu item! Perhaps it's the same for eagles!"

Tim then describes the 'caribou contact' with the calf photographed near Justin and Scott.

"How far were these animals when you obtained the first 'synchronizaTion'?" Alex asks.

"At least 500 metres," replies Tim. "It took about a half hour for the photo of the calf contact."

"That's about the same as our moose contacts where it is almost always a calf which firstly comes within a close distance." I comment. "How did your fox research go Joanna?"

"Great! The second den has at least seven kits and they play with juvenile snowshoe hares in a half acre of open forest grasslands. The adults seem to remain out hunting until dark so we had hours of progress. This could be a new record for a terrestrial mammal's rhythmic vocabulary."

"By the way Joanne's earlier described location, near the kittiwake colony is one of the finest places on Earth for an animal lovers paradise. And this includes humans!"

"After tomorrow, we'll try to get these results together for your analysis and comments. Soon, following the banquet, however, how does a Newfie beer in the bar sound. Then we can have a specially designed, multidimensional, multi denominational, after dinner discussion, relating to new concepts

---

[34] North Atlantic squid (Illex illecebrosus)

of a common spirituality, ideas mainly contributed over years by 'The Great Whales'."

Following this dinner celebration, we move into our charismatic, adjacent and photo decorated bar, where out come the single malt scotches plus a variety of cold beers for the eligible students.

CHAPTER 16

Surprising-Contact

We have all assembled in the partially darkened bar with our quality refreshments!

"Here is a thesis prepared for a large, soon to be held, academic meeting, so your comments and criticisms will be appreciated. I'll have the main points typed for distribution before any of our esteemed guests will have to leave tomorrow, late morning and I'll be traveling with this group. And Alex, we can even play my Tchaikovsky background music for your usually stimulating thoughts and comments."

"There are four phenomena in our world. A first is Mass, or matter. A second is Energy, which can, under certain circumstances, be related to Mass, by a value of the square of light velocity. A third is Information, based on: signals, signs, symbols, and 'Conventional time, t,' the latter being displacement divided by velocity, or simply space divided by speed. This is the temporal concept with which most humans are consciously familiar from an early age, as the faster one goes somewhere, the sooner one gets there! The fourth phenomenon is newly described as 'InformaTion,' which is based on a second temporal form called 'RhythmicTime, T' or 'Rhythm Based Time,' being a 'Mental Perception of Lateness Relative to a Biophysical SynchronizaTion!' 'Rhythm Based Time, RBT' is a foundation of 'Rhythm Based Communication, RBC,' or 'CommunicaTion,' which can be universal for living organisms and possibly also for unconscious entities. Based on the 'synchronizaTions' of planetary movements (such as sunrise and seasonal variations) all life has access to 'RBC.' We might recognize this form of 'communicaTion' from aspects of love and other positive emotions, but not from the many negative components of evolution."

"And this 'RhythmicTime, T, InformaTion,' not 'Conventional time, t, Information' using space divided by speed, could be considered to be a 'kingdom.' It is, to a large extent, a 'communicaTions kingdom' which we may more fully understand one day. And by using such 'Rhythmic InformaTion,' we may then learn efficient 'communicaTions,' using 'RBC,' with the rhythmic 'InformaTion' energies within Nature, including potentially those of deceased loved ones, presuming a concept of long and even everlasting life."

"Such a 'communicaTion' system both requires and produces low biological stress as we have discovered in its use by 'The Great Whales.' It requires low stress via 'synchronizaTions' but it produces low stress because only cyclical 'Rhythmic Time, T' matters, not signals, signs or symbols. If an animal scratches its ear 'onTime' it could have one meaning, at any other 'Time' of an agreed cycle, it could have no meaning. If a DNA molecule flexes 'offTime' it could have one meaning, at any other

'Time' of an agreed cycle, it could have another meaning, or no meaning at all."

"Unstressed Nature is closer to 'Rhythm Based Communication, RBC' or 'CommunicaTion,' and altruism, whereas stressed Nature is closer to 'Signal Based Communication, SBC' and evolution."

"Don't we find both in normal living?" asks Elliott.

"Yes indeed. But not in the elements of pure evolution, or its opposite, pure altruism."

MIND

"It is necessary to review where one's emotions, conscious and unconscious, lie relative to a modern definition of mind and its 'Mental Thought Processes, MTPs.' Human biophysics involves mind, which can also be called one's subjective reality. Scalar labels, such as colour, shape and entropy can be moved from one's objective reality by a mass and/or energy vector named a 'Mental Vector Process, or MVP.' One might remember this acronym as a 'Most Valuable Player,' but here, it is in the Game of Life, not necessarily hockey or sports!"

"There exist two forms of 'MVPs:' a, those from one's objective reality external to mind, such as the next serve in a tennis game and, b, those from internal memory storage, such as how the last game was won. These latter can originate in one's unconscious and emotions."

"Such subjective reality can also be called one's 'Essos,' standing for 'Event Space Sphere Or Spheroid,' pronounced 'Eee-sos,' which is a mental volume, with a 'non Essos' volume within, containing one's memory, knowledge, education, understanding, unconscious, emotions, cultures, etc."

"Also in review, mind or 'Essos' can be extremely dynamic, collapsing in volume during day dreaming, meditation and sleep. All of life has an 'Essos,' which is created by both external and internal 'MVPs'."

"A critical third aspect of mind is the organism's 'Mental Thought Process, MTP,' which receives 'Mental Vectors, MVPs' and then can produce bioscientific activity and/or biophysical action. It is here that aspects of most communications must originate in order to form such activity and/or action."

"Is this a unique, state of the art, say a working definition, of the human mind?" asks George.

"We believe so and Alex and others feel that the elements of this definition work for all living organisms."

"Emotions, which are normally not of mind, of 'Essos,' can attach concepts to internal 'Mental Vectors, MVPs' which concepts then become cargos transported to thoughts, 'MTPs.' The strengths and influences of such feelings can vary, normally, up to the causes and activity of actions; occasionally, however, they can be even greater."

"Let us consider how altruism could enter into 'Nature's RBC,' and how it could be contrary to evolution. An adult fox has been seen stationary on a grassy mound, observing a snowshoe hare some fifty metres distant. Providing the fox's needs are met, no hunger, etc., these animals could enter 'RBC' by agreeing upon any suitable high rise time, low stress communication signals, such as head and ear motions. Once synchronized, 'RBC' can commence; both fox and hare can be altruistic regarding an improved life style for the other. Such mutual altruism is a distinct survival mechanism for both parties and in this respect it can rival evolutionary theory. For example a hungry marine mammal could be altruistically told where food is abundant."

"Now the fox, which we'll say was 'communicaTing' about the full moon, gets hungry. This could be a feeling emanating from its unconscious. This is a form of biological stress and it disrupts the 'Rhythm Based Timing' of subsequent 'communicaTions' so that the snowshoe hare with rhythmic empathy feels the fox's hunger, quite possibly before the fox does! The empathetic and intelligent hare vanishes and lives. The less empathetic hare could perish, by evolutionary forces.

"The result appears to be that all organisms have emotional rhythmic empathy, for without such, their species would not have survived. In other words altruism and evolution can coexist, as long as 'Signal Based Communication, SBC' and 'Rhythm Based CommunicaTion, RBC' can both be a part of Nature."

"Such seems a logical proof of the existence of 'RBC.' Why then did we take so many centuries to discover it?" asks Alex.

"We had to wait for 'The Great Whales' to teach us Alex, and then perhaps that is a design that there is a spiritual caring for successful improvements for mankind."

'Coming Home Along Our Western Shore'

Sun sets behind a dark and wooded hill,
The sea is calm but dolphins flank our boat,
When suddenly sun rises with light to fill,
The nearby islands which show a silvery coat.

Sun sets again behind our rugged shore,
And ocean is quiet, except our dolphin friends.
But this time the rays of sun are shielded more.
Some think that surely daylight nearly ends.

But again sun rises, shining through a valley,
And again green islands glow in silver light.
For this time we shall not pause to dally,
As our homebound harbour isn't yet in sight.

Sun sets and rises, then sets again you see,
Because the shore does rise before it falls.
Such flickering effect we welcome with much glee,
The view of sea and sun is a voice that calls.

Many sunsets and sunrises angels send,
With the final landing by our home-built dock.
Then onward step toward our journey's end,
And the warmth of memories will for all unlock.

GLOSSARY

As Involved With 'RBC Theory And Experiments'

(Single quotation marks are used here and in the
text, primarily for new concepts.)

Altruism - Having regard for others; to give or to act
without reward; to be unselfish; such is contrary
to a survival of the fittest, in evolution.

Bidirectional - Functioning in two spacial
directions.

Biophysics - The science of the application of the
laws of physics to biological phenomena, most
especially involving organism mind.

'Cetacean-Contact' – Includes 'Rhythm Based
CommunicaTion, RBC' with whales, dolphins
and porpoises.

Clock - Any mechanism and/or life system that
represents, or is capable of producing cyclic,
recurrent or predictable motion, and measures
temporal qualities.

'ClosedWords' - Neglecting a normal space between words (e.g. OnTime). Used to signify the presence of 'Rhythm Based Time, T or RBT,' such is also called 'rhythmic Time.'

'Cluster of Temporal Scalar Labels' - A group of such labels existing within mind, until a change of thought of one's 'Mental Thought Process, MTP.'

Communication and 'CommunicaTion' - The passing of information involving: a) transmission, b) reception and c) the altering of subsequent behaviour. Please see 'SBC' and 'RBC.'

'CommunicaTion' - 'Rhythm Based CommunicaTion, RBC.'

Conventional - Traditional, in contrast or in opposition to the enclosed descriptions of rhythmic 'perceptions of lateness,' and/or 'Rhythm Based Time, RBT.'

'Conventional time, t,' or the commonly used noun time. Please see the following section named 'Conventional time Categories'.

Cyclic - Revolving in a recurrent series of events and/or phenomena.

Dimensions – 1. 'Spacial dimensions,' which define all known geometry, or 2. Variables, which may be scalar quantities, the semantic usage of which, as dimensions, is not recommended in order to avoid confusion with 1.

Displacement - Distance in a direction.

Duration - 'A conventional timetag of t,' of increased quantity less one of lesser quantity.

'DuraTion' - 'A rhythmic Timetag of T or RBT,' of increased quantity less one of lesser quantity.

'Earth Life' - An 'Earth source domain' using communication or 'communicaTion.'

Empathy - "Experiencing strong affection or passion" (Britain and U.S.), "Feeling into, as in watching a high-wire artist" (German), definitions by Dr. Frans De Waal.

'Essos' - 'Event Space Sphere Or Spheroid' (pronounced 'Eee-sos'). 1. 'A volume created by the mental analysis of received 'temporal scalar labels' and associated labels. 2. 'Essos' is a synonym for part of one's conscious mind where such part is an abstract volume useful to describe the dynamic orientation and magnitude of conscious space variables, real 'Now time and Now Time' and their combination called 'Now TIME.' 'Essos' and 'Ex-Essos' have been called "Subjective-inclusive Experience" and "Mind-independent Pre-structured Reality," respectively, by Dr. Herbert Muller of McGill University, Canada.

Future - Temporal, and other scalar labels, and their associated mass/energies (definitions to follow) that have not yet arrived at one's 'Essos.'

'Information (SBC)' - Information encoded in mass/ energy and using, for example, signals, signs and symbols, and described by conventional communication or 'Signal Based Communication, SBC.' ('SBC' definition to follow). Please also see 'World Phenomena' below.

'InformaTion (RBC)' - 'InformaTion' encoded in 'rhythmic Time T.' Such is potentially prolific in biophysics, and is found within perhaps all of Nature. This is the medium of 'Rhythm Based Communication, RBC' or 'communicaTion.' Please see definitions above and below as well as that of 'World Phenomena,' below.

Mass/Energy - Mass or energy or both.

'Mental Clock Process, MCP,' Found within a 'Mental Thought Process, MTP' and comprised of cyclical clocks, such as a suprachiasmatic nucleus, one or more of which clocks can be synchronized to the biological rhythms within the 'MTP' of an external mind, if an attempt at 'Rhythm Based Communication, RBC' is desired. Such 'synchronizaTion,' followed by a 'communicaTion' passkey, can help to lower the subsequent biological stresses, and thence move a following 'informaTion' exchange away from evolution, and toward altruism. Losing 'synchronizaTion' may then relate to a denial of such behavioural kindness.

'Mental Thought Process, MTP' - An area near 'Essos Centre,' as found for all of life, designated for the reception and processing of mind's 'Mental

Vector Processes, MVPs' (this very important definition follows directly).

'Mental Vector Process, MVP' - A mass/energy vector carrying a 'cargo' of one or more scalar labels from an 'Essos Edge,' to a 'Mental Thought Process, MTP.' ('MVPs' seem the 'Most Valuable Players' in the 'Game of Life.')

Mind (conscious) - 1. An organism's 'Essos' plus functioning memory. 2. Mass/energy and information, or 'informaTion,' involved with the architecture of a central nervous system and partially within one's 'Essos.' The important 'Essos' definition above is a synonym for part of one's conscious mind.

Mind (unconscious) - Mass/energy and information, or 'informaTion,' possibly involved with any active cell of a living being but not within 'Essos.' The important 'Essos' definition is above.

'Nature-Contact' - Includes 'Rhythm Based Communication, RBC' with humans and Nature.

'Nowness' - One's immediate present.

'Now time' - 'Conventional time, t,' associated with events in one's present and contained within one's 'Essos.' Please see 'Conventional time Categories' to follow.

'Now Time or NowTime' - 'Rhythm Based Time, T or RBT,' associated with events in one's present and contained within one's 'Essos.' Please see 'ClosedWords' above and 'Time and TIME Categories' to follow.

'Now TIME or NowTIME' – '1a. A mental perception of a Cluster of Temporal Scalar Labels and associated labels, and 1b, A subsequent mental analysis of such labels, from a 'Mental Thought Process, MTP, back to their sources.' Such an 'analysis direction' is opposite to the direction of the incoming 'Temporal Scalar Labels' and as such, it appears as a component of quantum mechanics. 2. 'Conventional time, t,' plus 'Rhythm Based Time, T or RBT,' associated with events in one's present and contained within one's 'Essos.' Please see 'ClosedWords' above and 'Time and TIME Categories' to follow.

'OnTime, LateTime, OffTime, EarlyTime' - Elementary cyclical 'Windows' of 'RBC.'

'Ontimeness' - synchronization (as defined below).

'OnTime' or 'OnTimeness' - An agreed, cyclic, reference 'Time T' for 'Rhythm Based Communication, RBC, or CommunicaTion.'

Orthogonal - At right angles or 90 degrees.

'Orthogonal Spacial Dimensions, OSDs' - There are a maximum of only three. (as an example: east, north and up). Please see 'Space' definition below.

Paradigm - A mode of viewing the world which underlies scientific theory for a period of history.

Paradigm Shift - A fundamental change in approach and/or philosophy.

Past – Temporal, and other scalar labels, and their associated mass/energies that have previously departed from one's 'Essos.'

Perception - An interpretation based on one's understanding.

'RBC' - Please see 'Rhythm Based Communication,' plus Communication vs. 'CommunicaTion.'

'Rhythm Based Communication, RBC or CommunicaTion' - 1. 'A form of disclosure encoded in rhythmic Time T,' which can be experienced in certain forms of human feelings, and is potentially practiced within much, or perhaps all, of Nature. 2. Encoding in 'RBT' and using 'InformaTion (RBC) or RBI.' Please see definition of 'Rhythm Based Information, RBI.'

'RBI' - Please see 'Rhythm Based Information, Information (RBC) and World Phenomena.'

'Rhythm Based Information, RBI, or InformaTion' - 'InformaTion' encoded in 'rhythmic Time, T or RBT,' which is potentially prolific in biophysics, and is found within perhaps all of Nature. Please see 'World Phenomena.'

'RBT' - Please see 'Rhythm Based Time' or 'T'

'Rhythm Based Time, T or RBT' (as opposed to 'Conventional time, t). - 'A mental perception of lateness relative to an agreed, biophysical, cyclical concept of rhythmic synchronization, including 'onTimeness,' between two or more minds.' Please see 'Time and TIME Categories'

'Rhythmic Time or RhythmicTime, T or RT' - As 'Rhythm Based Time, RBT' (above).

'SBC' - Please see 'Signal Based Communication.'

'Signal Based Communication, SBC' - 1. 'A form of disclosure encoded, for example, in signals, signs and symbols, and using conventional, temporal, scalar labels.' 2. 'Signal Based Information, SBI' encoded in mass/energy. Please see definition of 'Signal Based Information, SBI.'

'SBI' - Please see 'Signal Based Information, Information (SBC) and World Phenomena.'

'Signal Based Information, SBI' - 'Information encoded in mass/energy and using, for example, signals, signs and symbols.' Please see 'Information (SBC) and World Phenomena.'

Scalar - Having only magnitude, without spacial direction, that is without spacial dimensions.

Space – Can be described by orthogonal vectors, always containing both magnitudes, and spacial directions of: 1. right-left, forward-backward, up-down, 2. east-west, north-south, up-down, or 3. north celestial pole, declination, right ascension east of the 1st Pt. of Aries. 1. is subjective or 'of Essos,' 2. and 3. are objective.

'Spacial Dimensions' - Vectors which are most often considered orthogonal (in which case there are up to three and only three spacial dimensions). Please see definition of 'Space,' above.

Spacial Directions - Space vectors, which are not necessarily orthogonal.

Symmetric - When certain positions rotate into other positions in the same set.

Sync - An abbreviation for either synchronization or 'synchronizaTion.'

Synchronization - Happening at the same linear OR cyclical, 'Conventional time, t.'

'SynchronizaTion' - 1. Happening at the same cyclical, 'Rhythm Based Time, T or RBT' (also 'Rhythmic Time, T'). Such is also 'rhythmic synchronization.'

'Temporal Scalar Labels' - Are those that leave a clock as a cargo of a mass/energy vector, enter a 'Mind' as real 'Conventional time' and/or real 'rhythmic Time,' and eventually may become scalar 'conventional timetags' and/or scalar 'rhythmic Timetags' (see below).

'Temporal Tags' - Please see 'Temporal Scalar Labels' (above), and 'Conventional timetags, Timetags, and TIMEtags,' to follow.

'True altruism' - To give or to act, as in altruism, without the expectation of a reward.

Vector - A quantity having spacial direction as well as magnitude.

'Whale-Contact' - Includes 'Rhythm Based Communication, RBC' between humans and whales.

'Windows' - Relatively short 'RBT duraTions' of 'Rhythm Based Communication, RBC.'

'World Phenomena' – 1. Mass, 2. Energy, 3. 'Signal Based Information in t,' 4. 'Rhythm Based Information in T.'

'Conventional time Categories'

'Conventional communication' - Please see 'Signal Based Communication, SBC.'

'Conventional time, t,' or the commonly used noun time. 1. 'A mental reading OF a clock,' not a reading ON, AT or IN a clock. (Please see definition of 'mind.) 2. Displacement divided by velocity or space divided by speed, as a 'temporal, scalar label' which can move in the direction of its associated mass/energy vector. Traditional concepts of time differ from the new 'Rhythm

Based Time, T or RBT.' Please see 'Time and TIME Categories,' to follow.

'Conventional time, t, tags' - The association, either physical or mental, of 'Conventional time, t' scalar labels, with mass/energy vectors, or simply 'time t' scalars. Please see following three definitions.

'Conventional timetags (developing)' - Varying temporal scalar labels of 'Conventional time, t,' produced by a working, linear or cyclical clock.

'Conventional timetags (fixed)' - Temporal scalar labels either mental, inscribed, machine made, geophysical, geologic or others, of 'Conventional time, t.'

'Conventional timetag-vector' - A 'conventional timetag,' on a mass/energy vector. Some call this a time-vector, which is discouraged as 'Conventional time' does not flow on its own.

'Time and TIME Categories'

'Rhythmic Timetags (developing)' - Varying 'temporal scalar labels' of 'Rhythmic Time T,' produced by a working cyclical clock.

'Rhythmic Timetags (fixed)' – 'Temporal scalar labels' either mental, inscribed, machine made, geophysical, geologic or others, of 'Rhythmic Time, T.'

'Rhythmic Timetag-vector' - 'A Rhythmic Timetag on a mass/energy vector.' Some call this a Time-vector which is discouraged as 'Rhythmic Time' does not flow on its own.

'T' - Please see the following definition of the new 'Time' or 'Rhythm Based Time, RBT.'

'Time,' 'Rhythm Based Time, T or RBT.' – 1. 'A mental perception of lateness relative to an agreed, biophysical, cyclical concept of rhythmic synchronization, including 'onTimeness,' between two or more minds.' 2. Via Dr. Hitoshi Kitada (Tokyo): "T=exp(it(+/- 2pi)H/Planck constant h), and for any complex number exp(i theta) on a sphere of radius one (where theta is any fixed real number), then exp(i theta) exp(it(2pi)H/h is a solution of the Schrodinger equation where H = a Hamiltonian operator, and for theta = (2n+1)pi, (and n = 0, +/- 1, +/- 2, - -), exp(i theta) = -1 and T becomes both bidirectional and "rotation free."

'TIME' - 'Conventional time, t,' plus 'Rhythm Based Time, T or RBT,' within 'Essos.'

'Timetags' - See 'Rhythmic Timetags' as above (2 types).

APPENDIX 1

'RBC' and General Experimental Technique

(The scene is the early morning student meeting
at The Village Inn, during Chapter 1.)

"Firstly choose a time of low internally and
externally caused stress, such as, no need for
nutrition or, no fear of mind/body discomfort.
Then 'share a rhythm!' This can be interpreted
as energy packages traveling between organisms,
which signals have a common, compatible,
between pulse 'duraTion' called an 'alpha concept.'
Upon synchronization, these 'duraTions' become a
shared rhythm. Now add a second simple message
called a 'beta concept,' as explained presently, and
to your surprise, 'Rhythm Based Communication,
RBC' can begin!"

"It might be useful if you gave common examples
of' alpha concepts' and some good communication
signals for both whales and land mammals as well
as bald eagles," interjects Nicholas.

"OK, or as some of you would say now, just
'kay.' For roosting eagles we use an alpha concept
or rhythm of 12 seconds with mainly whistles and
light flashes for signals. For fox kits and snowshoe
hares we use a rhythm of 20 seconds with mainly

finger snaps and rock taps. For beaver, black bear, caribou and moose we use 32 seconds, underwater sounds for beaver but mainly comb flicks and light flashes for bear, caribou and moose. For whales we use 60 and 90 second 'alpha concepts' with computerized, underwater, acoustic transmissions of various intensities. By the way Nick and Kirk, choose four to join *Ceres* this morning on the water. Everyone should get chances when whale contact space is available.

"Next you transmit non frightening, rhythmic, 'alpha concepts,' called 'alpha rhythms'.

"Record identification and behavioural details of any animal that returns a signal appearing to be at the same time as your transmitted signal, which behaviour is 'possible-synchronization,' or 'possible sync.' Cease all transmissions for a 10-15 minute coffee break if no sign of a 'possible sync.' occurs in 15-20 minutes. Postpone the experiment if active feeding or external stress is suggested.

"Following 'possible sync.,' wait one 'alpha concept' and then send the next signal 'late,' which is a 'beta concept.' This is 'the first message signal.' Transmit another signal 'late' using the identical amount of 'lateness.' This is 'the second message signal.' It will and must follow 'the first message signal' by the 'alpha concept,' which, in other words, is the 'pulse interval' before synchronization.

"Record all signals from your single animal or any in a group but place maximum effort in following and observing the animal associated with the suspected synchronization. Note that you have not proven anything beyond coincidence yet, but, that will come next, as you hopefully enter the world of 'RBC.'

"Transmit the third message signal which must be 'onTime' or at a 'RhythmicTime' corresponding to the beginning of the synchronized 'duraTion' which is also called the 'alpha time.' Successful 'RBC' is now, complete if receiving partial message mimicry, so keep up a maximum effort to observe and record, without transmissions.

"For example, a transmitted message of 'late, late, onTime,' which gives a 'key' of the amount of lateness, or the 'beta concept,' should return the key with a mutual understanding of both the alpha and beta concepts, when you receive signals which are 'late, late, onTime.' Then, return to the main research lab, across Taverner's Path from the Inn, or find Tim, Kirk, or myself with news of success, good data or new knowledge."

There is student silence, save for the scratching of note taking. Then abruptly hands go up.

"Exactly what is an 'alpha concept'?" asks one.

"It is an idea in your mind, presumably transmitted to the mind of an organism, which is simply the between pulse 'duraTion' during intended 'synchronizaTions.' We use this 'duraTion,' 'synchronizaTion' and 'onTime' etc. spelling to emphasize the 'RhythmicTime, T,' and not 'Conventional time t'," as I copy these 'RBC' words onto a flip chart.

"Then what again is a 'beta concept'?" questions another.

"It is an idea in your mind, hopefully transmitted to the mind of an organism, which is simply the 'duraTion' of the first one quarter of the alpha concept, rhythmic cycle. It starts with 'synchronizaTion' and ends in the 'lateTime' window. Similarly, the 'gamma and delta concepts' are the 'duraTions' of half and three quarters of the same rhythmic cycle, ending in the 'offTime' and 'earlyTime' windows. respectively."

Tim joins in with, "Your opening message, now called a 'passkey,' or in human concepts simply a greeting, will then be as mentioned, 'late, late, onTime,' or 'beta, beta, alpha.' This has already been programmed into your computers by hitting both the 'Psion' and 'P' buttons at the same time."

"Humpback whales are now teaching us that 'offTime, offTime, onTime' or 'gamma, gamma, alpha,' is another and different passkey. This message seems more universal, so you might want to try it if you don't get message mimicry from the programmed passkey. I'm beginning to feel that both keys work but convey slightly different but as yet unknown specific meanings."

APPENDIX II

21st Century Biophysics of "Two Temporal Types"

'Conventional time, t' is unidirectional as space divided by speed. 'Rhythm Based Time, T' being bidirectional is 'A Mental Perception Of Lateness Relative To An Agreed, Biophysical, Cyclical Concept Of Rhythmic Synchronization (Or SynchronizaTion) Between Two Or More Minds.' For circular motion, where 2 pi radians = 360 degrees, t counts 0, 2pi, 4pi - - -, whereas 'T' counts 0, 2pi = 0, 2pi = 0.

In biophysics, because of the workings of all living organism minds, (t + 'T') is mentally acceptable but (t x 'T') is not. Example: travel 12 hours south from Niagara Falls in t, and in the same 12 hours Lake Ontario space has rotated to approximately Lake Baikal space, in Russia, in 'T.' Adding these temporal concepts (12 hrs + 12 hrs), or displacement concepts gets you to 'China space,' but multiplying (12 x 12) is mentally meaningless.

1) The Lorentz Transformation (reference: 'Relativity, by A. Einstein, equation 11a, page 137, term #4, involves t of light velocity x 'T' (which here is designated t-primed of system K-primed, with "axes pointing in mutually exclusive directions"). Thus it is believed that (t x 'T') in not mentally acceptable.

2) Both temporal types are scalar labels, which biophysically should not be combined with vectors, such as the three and only three 'Orthogonal Spacial Dimensions' (e.g. east, north and up). Thus 'space-time' (as 'space-colour') is a non sequitur, which could be 'time-in-space' (or 'colour-in-space'). If space is curved then time-in-space is not necessarily curved, because of the biophysical concept of: 'NO MIND, THUS NO TIME.' Furthermore, gravity could very well now be caused by impacting particles, as for example neutrinos.

3) The 'Twin Paradox' is not a paradox any longer because it involves (t x 'T'). One's age is measured in 'Rhythm Based Time, T,' the rhythm of the Earth's motions.

References:
Einstein, A., Relativity, 188pp, Three Rivers Press, New York, 1961, ISBN 0-517-88441-0
Einstein, A., Collected Papers, Princeton University Press, 2008, ISBN 0-691-08526-9
Sartori, L., Understanding Relativity, University of California Press, ISBN 0-520-20029-2

BIOphysics - versus - 20th Century Physics

A great scientist's fine accomplishment is that energy is proportional to mass.

On the other hand, the 'Lorentz Transformation,' Can only work if there exist no 'Rhythm Based Velocities.'

For 'Rhythm Based Time, T,' cannot form a
quotient,
With the 'Conventional time, t,' of light velocity.

And, unfortunately, 'space-time' is a non sequitur!
Space is a vector with magnitude and direction,
But time is a scalar with magnitude only.
Moving vectors can transport scalars but they
RADICALLY differ.
Biophysics must replace 'space-time' with
'time-in-space.'
And so curvature of space does not infer
curvature of time.

Additionally the 'Twin Paradox' is not a paradox at
all.
One twin travels in 'Conventional time, t,'
But ages are in 'Rhythm Based Time, T.'
And one must not multiply these two together.
Therefore, both twins will always remain as twins,
As long as the 'Rhythm Based Time' of Earth stays
the same.

Still, a great scientist's fine accomplishment is
that energy is proportional to mass.

APPENDIX III

'Now TIME' seminar continued -- 'ALL (real) TIME IS NOW TIME'

'Conventional time, t' is 'a mental reading OF a clock,' not a reading ON, AT or IN a clock.

'Rhythmic Time, T' is 'a mental perception of lateness relative to an agreed, biophysical, cyclical concept of rhythmic synchronization, including onTimeness, between two or more minds.'

'All (real) TIME (t + T) is Now TIME' as '1. a mental perception of a Cluster of Temporal Scalar Labels and associated labels, and 2. a subsequent mental analysis of such labels, from a 'Mental Thought Process, MTP,' back to their sources.' Such an 'analysis direction' is opposite to the direction of the incoming 'Temporal Scalar Labels,' and as such, it appears as a component of quantum mechanics.

"One's 'Now TIME,' where 'TIME' = t + 'T,' is related to the mean radius of one's 'Event Space Sphere Or Spheroid (Essos),' which space may have a variety of shapes. To convert this radius into a measure of 'Now TIME,' for example, seconds or microseconds, divide by the average velocity of the energy received by one's 'MTP,' near 'Essos Centre.' One's future, lying outside such an Event Space, is comprised of mental, inscribed, or machine made invented labels, which we have called 'timetags.' and 'Timetags.' Similarly, one's past is comprised of, mental, inscribed, or machine made 'time/ Timetags,' which may be real. But these labels, real or invented, are not real 'TIME' unless they lie within one's 'Essos.' And thus is derived the important concept that:

'ALL (real) TIME IS NOW TIME.'

Additionally all (real) 'TIME' has been defined as: 'The Reading ON a Clock,' whereas it should be "The Mental Reading OF a Clock."

"These 'magic words' seem to be a key to the use of a new communications logic to begin the vital process of merging our human species, altruistically, with an 'Earth entity' that many now perceive and simply call, Nature, and also some now call it Gaia."

"'Now TIME' has not in the past been a part of the 'language of physics,' but it critically needs to be accepted as important for an attempted amalgamation of biology with all other sciences, for the forthcoming temporal description. This point will now be expanded in ten ways."

"1. 'Conventional time, t' and 'Rhythmic Time, T' are scalars, can be 'counting scalars,' but they are NOT vectors and are certainly not 'arrows'."

"A 'counting scalar' is a quantity with magnitude only which changes approximately linearly, as does the number of rotations of the Earth about its polar axes. In marked contrast a vector has not only magnitude but also direction. An 'arrow of time' is Eddington's misleading metaphor, for time has no direction (no north, no south, no up, no down) and no target could ever be hit by such an arrow!"

"Also, try to imagine how you could ever construct a bow to fire such an arrow," adds Nicholas.

"Exactly! Unreasonable! Impossible!

"2. All real temporal concepts are 'Now time t, Now Time T, or Now TIME,' where 'TIME = t + T'."

"I believe that this is one of the most important biophysical and psychophysical concepts of all of Nature and especially of our human species. It, at first, seems to be joining subjective, cerebral time to a traditional past-to-future, objective 'arrow-like' quantity. Thus, it may be a difficult concept to understand, for someone trained and specializing in just theoretical physics; but other compatible components of a proposed fuller meaning of time, hopefully a meaning even more accepting to physicists as well as to other scientists, follows directly."

"So, you will soon give everyone a working definition of 'Now TIME,' I presume!" states Alex, who has just entered the 'Eagle Room,' and is seated as a nearby table, with his large coffee mug.

"Certainly, but we must firstly define 'timetags and 'Timetags,' and our proposed dimensions of the 'Event Space' concept, and, we must talk about past time and future time."

"3. Past time involves 'timetags,' or time labels, which may consist of factual sensory signifiers."

And what on earth does he mean by 'timetags' think some? Employing empathy, this feels like a question on many minds.

"'Timetags and timetags' are mental, inscribed or machine markers, nothing more."

"Can you give us factual cases of these categories?" asks Mark.

"Certainly, examples of past 'timetags' are incorporated in: the recollection of where you were on 9/11 or when President Kennedy died, an inscribed date on the cornerstone of a building, or yesterday's 'Rhythm Based Time' data about to come off our computer printers upstairs."

"4. Future time involves 'timetags' consisting of invented sensory signifiers.

"Examples of future timetags are incorporated in 'dreams of a better tomorrow,' plans for today's experiments, or imagined tastes of our joint lobster banquet at the end of this special expedition! Future time and past time are thus described to be somewhat similar, but, they are bridged by the anomalous 'Now TIME,' which may act as a potential 'hinge' of symmetry."

"Hold it please! What do you mean by a hinge of symmetry?" asks Charles.

"Well, the symmetry in this case is an arrangement of 'timetags' and 'Timetags' in similar, opposing positions, with respect to a geometric plane of division. And this is a very specific plane, through an 'Essos,' orthogonal to incoming 'Mental Vector Processes, MVPs' and intersecting a 'Mental Thought Process, MTP'." Please see Glossary.

"So where does the hinge come in?" responds Charles.

"It is as a figure of speech representing the folding or hinging of a small duration in one's future, into a duration of the same apparent length, conceived as in one's past."

"5. 'Conventional time, t' is the result of 'temporal scalar labels' which become attached to a 'mass/energy-vector', creating moving 'timetags.' Such projections can create both past or future 'timetags.' The equations of Special and General Relativity are, to the state of present knowledge, satisfied by the resulting 'timetag-vectors,' except when they mistakenly include a quotient of t and 'T'."

"A 'timetag-vector' which is an important new concept is 'the direct association of temporal, mental, inscription or machine markers with any mass and/or energy (also mass/energy),' even should such quantities exist with zero relative motion. Recall that if an object in Nature seems apparently motionless relative to the Earth's surface, it still is a mass-vector, relative to an external coordinate system, due partially to the daily (diurnal) rotation of the Earth. Living objects continually exhibit many 'mass/energy-vectors,' any one of which may carry 'timetags,' by this mental, inscription or machine process."

"But diurnal, as opposed to nocturnal means during the day vs. during the night!" says Tim.

"You are correct as for one of the meanings. We use the astronomical meaning, however, which is approximately 24 hours or 1440 minutes."

"6. 'Rhythm Based Time, T or RBT' is the result of temporal scalar labels which are 'projected,' by human mind, inscription or machine, as 'Timetags' (as 'LateTime, OffTime, EarlyTime, OnTime,' etc.) onto a 'mass/energy' or conceptual vector and are so attached only in an 'RBT Plane,' which can be orthogonal (at right angles) to the above mentioned principal axis of the said vector."

"'RBT' can be defined as: 'a mental perception of lateness relative to an agreed, biophysical, cyclical concept of synchronizaTion, including onTimeness, between two or more minds.' Our name for such rhythmic synchronization is 'OnTimeness' spelt with an upper case T, only to differentiate it from 'Conventional time, t.' 'RBT' is an encoding mechanism for 'Rhythm Based CommunicaTion'."

"So 'RBT'measures particle spin," states Tim.

"Yes! And hence it has two directions. We will deduce, momentarily, that 'RBT' is indeed bidirectional."

"7. 'Conventional time, t,' appears to move in the direction of its 'mass/energy vector' and with few exceptions is considered, by convention, to always move toward 'the future'."

"Physics has never before been able to answer the question: How fast does 'Now TIME' move? If we use the fourth concept as previously stated, that 'future time' involves extrapolated 'timetags,' then we all surely agree on a highly probable prediction that 'Now TIME' will move along the principal axis of a mass/energy physics vector, to the 24th hourly 'timetag,' in one diurnal rhythmic cycle of 'RBT.' In other words, the clock for the 'apparent progress' of time in physics, actually exists in the union of: a, geophysics, b, biophysical 'Now TIME' and, c, 'Time/ timetags'."

"But if 'Conventional time' can 'flow,' then the arrow metaphor should work!" comments Justin.

"No! Temporal concepts, in fact, do not 'flow;' they are counting scalar quantities with no spacial dimensions! The 'apparent progress,' as just mentioned, is a product of the fact that classical time t, by convention, always counts toward 'the future' and, that 'timetags' are 'attached' to mass/ energy which does move in a particular spacial dimension."

"8. 'RBT' is bi-directional (say clockwise or counterclockwise) and may be associated with 'vector spin,' including rhythms, and especially biological rhythms such as heart motion, breathing and 'body language.' It can also be associated with geophysical rhythms such as the diurnal (perhaps the most important of all life rhythms), the lunar and the annual."

"Can you give us an example of a lunar rhythm in biology?" asks Edward.

"Marine intertidal organisms surely comprehend the lunar rhythms, which cycles, after all, control the tidal oscillations."

"9. 'RBT' is used as a communication encoding mechanism for 'low stress' organisms and as a result it seems an ideal candidate for a more prolific information flow throughout 'low stress' Nature."

"And that is our major discovery!" adds Alex.

"10. So let us define a new 'TIME' by the equation: 'TIME = t + RBT' or, 't + T' where t and 'T' are 'timetag-vectors.' Because of the cyclical nature of 'T,' 'TIME' will thus also be associated with 'vector spin.' The above resultant sum represents a 'helical information package,' perhaps leading to new inquiries in the direction of recent discoveries in the fields of molecular biology and biochemistry."

"You are referring to RNA and DNA, I presume?" inquires Nickolas.

"Exactly, and more!"

Then Alex approaches the group and while standing adds, "Several of these items are changes to human concepts of the logic of time but it appears suggestive to some physicists and biophysicists that they are at least a foundation, part of the picture puzzle; but that there is room for added pieces of the logic in order to, one day, finish a more comprehensive, if not complete, understanding of time."

"The true temporal logic would seem to some, however, to be one of the great mysteries of modern science, and a reason may well be that we need the above mentioned amalgamation of biology with physics."

APPENDIX IV

'Signal Based Information' versus 'Rhythm Based Information'

Real 'TIME' is 'The Mental Reading OF A Clock.' Said clock is either: a) 'body-external' (geophysical, biological, or man-made), or b) 'body-internal' (biophysical and biochemical). All other known temporal concepts are scalar labels called 'timetags of Conventional Newton/Einstein time, t,' or, 'Timetags of Rhythm Based Time, T.' If such scalar labels, while transported by mass and/or energy vectors, enter a mental 'Event space sphere or spheroid (Essos),' they change from scalar labels (such as those ON a clock), to real mental readings of 'TIME,' where this upper case 'TIME' is ('t + T'), existing only within 'Essos,' within mind. Therefore it should become clear that 'If No Mind, Then No TIME.' Temporal concepts leaving a mind can remain as scalar labels. What causes such scalar labels to change into real 'TIME' is that one's 'Mental Thought Process, MTP,' analyzes the incoming vectors backwards in their 'conventional time, t,' and forthwith delivers a 'real TIME' mental picture to active thought, as a viable concept of 'Now TIME.'

Information (of one type), is encoded in mass and/or energy (signals, signs, symbols) and is so named 'Signal Based Information.' However, a newly discovered 'InformaTion' (of another type) is encoded in 'Rhythm Based Time, T,' which 'Time' is hereby described as 'a mental perception of lateness relative to an agreed, biophysical, cyclical concept of rhythmic synchronization, including onTimeness, between two or more minds.' Both 'Signal Based Information,' and 'Rhythm Based Information,' are transmitted in 'Conventional Newton/Einstein time, t,' (also, scalar space/speed). 'Signal Based Information' appears to form a 'foundation' of Evolutionary Theory. 'Rhythm Based Information' appears to form a 'foundation' of Altruistic Theory. The new 'Rhythm Based Time, T,' leads to Nature's 'Rhythm Based Communication,' abbreviated as 'CommunicaTion.' And most importantly, this novel 'CommunicaTion' is independent of both specific mass and specific energy, and may thus be related to feelings, sensitivity and spirituality.

BIOPHYSICS
VERSUS
TEMPORAL HISTORY

A great scientist produced many copies of
temporal history.
He and others considered an 'understanding' of
'the mind of God!'
But if God is 'InformaTion' {encoded in 'Rhythm
Based Time, T' (1), and not 'Conventional time, t'},
then such a God has no human-type mind,
no 'CommunicaTion Boundaries' (2),
and no 'Essos' (3).
And such a 'CareTakerGod' has a house with
many, 'many mansions' (John 14, 2).

And, stated is, that a theory 'exists only in our
minds!' Au contraire, as memory and knowledge
etc., need not be a part of one's 'Mental Thought
Process, MTP,' or within one's 'Essos' (3), one's
'Now TIME,' (4), or one's Human Mind (5). Recall
that 'TIME = t + T.'

And, a mentally received event is something
that takes place AT (or before) an 'Essos Edge,'
producing the important 'Mental Vector Processes,
MVPs' (6), which are vectors transporting scalar

labels that carry event characteristics to a 'Mental Thought Process, MTP,' the latter being located approximately at 'Essos Centre.'

A scalar label of event timing (a 'timetag') is a reading ON a clock, but a 'timetag' that arrives at a 'Mental Thought Process, MTP,' is the reading OF a clock. The latter can be 'real time,' in fact it can be real 'now time.'

Similarly a scalar label of event 'Timing,' that is a measure of event 'Rhythm Based Time, RBT or T' (a 'Timetag'), is: a reading ON a clock, and a 'Timetag' that arrives at an 'MTP' is a reading OF a clock and can be 'realTime,' in fact it can be real 'NowTime.' (Closed words additionally signify 'RBT, T' or, 'RhythmicTime.')

(1) 'Rhythm Based Time T' is: 'The Perception of Lateness Relative to a Common SynchronizaTion Between Two or More Minds.' (We use 'synchronization' when describing 'Conventional time, t or Ct.')

(2) 'CommunicaTion' is solely that which involves 'RhythmicTime, T,' and not 'Conventional time, t.'

(3) 'Essos' is an acronym for 'Event Space Sphere Or Spheroid,' (pronounced 'Eee-sos,') which is a <u>dynamic</u> volume surrounding a 'Mental Thought Process, MTP,' but not including an inner volume containing memory, knowledge, understanding, education, culture, and one's unconscious concepts.

(4) 'Now TIME,' consists of present temporal scalar labels ('timetags' and 'Timetags') within 'Essos.'

(5) A Human Mind has: a, Two 'Essos Edges,' b, 'Mental Vector Processes, MVPs,' and c, the receiving mechanisms for 'MVPs' which have been entitled 'Mental Thought Processes, MTPs.'

(6) 'Mental Vector Processes, MVPs' are mass and/or energy vectors which obtain 'cargos' of scalar labels (time, colour, shape, etc.) mainly at 'Essos Edges,' and deliver such information to 'MTPs.'

This important acronym, 'MVP,' seems the 'Most Valuable Player' in the Game of Life!

A great scientist produced many copies of temporal history, now we must consider 'TIME's FUTURE.'

POSTSCRIPT

"Here is a seemingly logical and novel merging of 21st century science with spirituality! This new concept of 'InformaTion' (recorded from unstressed animals, as in this book and in its earlier sequel 'Dancing With Whales,' as well as journal publications) leads to a form of 'communicaTion' encoded in <u>'Rhythm Based Time, T or RBT.'</u> Throughout this research we use both an internal upper case 'T' and closed words (such as 'onTime') to designate this newly discovered temporal form, as described by: <u>'A mental perception of lateness relative to an agreed, biophysical, cyclical concept of synchronizaTion, including onTimeness, between two or more minds'</u> (measured in angular displacement divided by angular velocity, please see Glossary)."

Having listened carefully to Peter, Hans poses a question.

"Could it not be that this 'informaTion kingdom' is a 'Kingdom of Heaven' contained within our biosphere? And if so where is it?"

"Wow!" says Timothy.

"Yes! You've hit the nail on the head and the location of such is not in inanimate matter, for therein lie no beneficial synchronizations. Neither is it presumably in lifeless voids or in 'ultra fine particles.' Such a kingdom is mainly one of 'Rhythm Based InformaTion' between cellular structures containing helical macromolecules, as found in all life and which are prolific in the human conscious and unconscious minds. An obvious starting 'synchronizaTion' is the Earth's daily rotation, with its diurnal rhythm, thence the moon which leads us to tidal rhythms. Other 'synchronizaTions' can involve verbal, behavioural or electronic 'communicaTions,' weather systems, or even seismological activity. Such a kingdom is invisable, coming not with potential visual observations because it is 'informaTion' of 'RhythmicTime,' and not information of mass and energy.

"An atheist has no knowledge of 'Rhythm Based CommunicaTion, RBC,' and thus many long lived organisms, perhaps long because of 'RBC' and its low stress, probably consider atheism to be incomprehensible. Agnostics may have feelings of a deity, but no conceived logic of 'Rhythm Based CommunicaTion,' so they presume that scientific progress re spirituality does not seem worthy of pursuit. A prescription for these folks, for a new sense of belief, could be to observe for a while, the 'communicaTions' with and between 'The Great Whales'."

"Christianity has a rhythm of seven days whereas Islam shortens this to several hours. Buddhism's mantras can have durations of seconds but they can also have lengthy intermittences requiring re-synchronization. Meditation does create low stress, possibly because of internal, unconscious 'synchronizaTions,' which in the case of Buddha, may have transferred to direct mental information.

So also, the later life of Jesus indicates similar direct mental 'CareTaker Contact' leading him to claim a 'jewel' of Christianity, namely that "THE KINGDOM OF HEAVEN IS WITHIN US" (Luke 17, 21, et al.). Such a kingdom would thus seem to involve internal rhythms, presumably for all living organisms. Additionally 'The CareTaker' has a "HOUSE WITH MANY MANSIONS" (John 14, 2), in which case such a 'house' could be within every living set of 'communicaTion' macromolecules evolved via all life on planet Earth."

"If the 'Kingdom of Heaven' is within or even amongst us, then such a kingdom appears to be 'RhythmicTime InformaTion.' This may lead to 'A Form of Everlasting Life,' with no evolution and no negative emotions. In such a case, 'A Life' as we commonly describe it, could consist of 'SBC (Signal Based Communication)' and 'RBC' with mixed emotions, and 'Heaven' could consist of 'pure RBC' with positive emotions."

'The CareTaker' can be described as 'A Universal Deity Comprised of VAST, VAST Amounts of Rhythm Based InformaTion.' In contrast, a component, entitled 'Our CareTaker' (also potentially a care giver) is hereby suggested for humans only. Some of 'Our CareTaker's Rhythms' can have 'synchronizaTion' with the biophysical rhythms within each and every living human (potentially the macromolecules of one's conscious and unconscious minds), based on the Christian prophesies of Luke 17, 21, 1 Corinthians 6, 19, et al., which can be translated as: (part of) 'The Kingdom of Our CareTaker is Within Every Human;' and also, (part of) 'Such a Kingdom is Amongst Any Gathering of Humans.' By analogy one could thence presume that (part of) 'The Kingdom of The CareTaker,' Nirvana or 'The Kingdom of Heaven,' is potentially within every living being. Recall that these 'Kingdoms' are hereby defined as 'Rhythm Based CommunicaTion SysTems,' based solely on, and encoded by, the novel temporal form, 'T or RBT,' as defined above."

"Such 'synchronizaTions' are 'passkeys' to 'Rhythm Based CommunicaTions, RBCs' (or more simply, we use the concept of 'rhythmic communicaTions' or, a single-word written concept of 'communicaTions'). Such are encoded in 'Time (RBT)' and NOT the communications that we so commonly use, which are encoded in signals (as language), signs, symbols and 'conventional time, t' (displacement divided by linear velocity, or space divided by linear speed)."

"Conventional synchronization means happening at the same 'conventional time, t,' which always, by definition, advances into one's future. 'Rhythm Based SynchronizaTion' (or more simply, we use the concept of 'rhythmic synchronizaTion' or, the single-word written concept of 'synchronizaTion') means happening at the same 'Rhythm Based Time, T,' which, by definition, remains as a component of one's 'NowTime' (for 'Now, NowTime and NowTIME,' please see Glossary)."

"'The CareTaker' has an upper case 'C' for dignity and an upper case 'T' to represent the new 'Rhythm Based Time, T.' Such a concept can be further dignified as 'The CareTakerGod'."

"From the beginning of life, via 'Rhythm Based CommunicaTion, RBC,' let us presume that 'informaTion' (not 'Signal Based Information') began to accumulate in a 'Kingdom of Heaven,' whereby its genderless landlord and Master, 'The

CareTakerGod,' became omnipotent, omniscient and omnipresent, in such a 'Kingdom' via the use of 'Rhythm Based InformaTion, RBI.'

"'RBI' has <u>no designated mass and no designated energy</u> which could account for the fact that scientific methods such as recording and photography have not verified the presence or any physical nature of 'Our CareTakerGod.' As stated in Luke 17, 20 'The kingdom of God does not come with observation.' However, universal spirituality assumes the existence of a deity, as one's emotions might tell us that we appear 'cared for' both in this life and onwards. Such seems valid through the reception of 'Rhythm Based CommunicaTion' and 'Specific InformaTion' as well as a history of <u>spectacular prophets</u> who have translated such 'InformaTion.' As said the prophet Gautama Buddha: 'To realize nirvana, this is the blessing supreme' (Mangala Sutta)."

"Yes, these discoveries, based on <u>'Rhythm Based InformaTion's potential, spiritual movement into our understanding of Nature,'</u> predict a long or even an 'everlasting' life."

POSTSCRIPT POEM

"THE MUSIC OF NATURE" and THE NATURE OF "RBC"

There is underlying rhythm in places of places,
There are underlying notes of 'ecofaces' in faces,
Such a law so well known, such that all of Earth
changes,
But there's change not yet dreamt in where all
rearranges.

For the Music of Nature isn't just in the timing,
As this music isn't rhythm, this music isn't
rhyming.
For the rhyming is signal and rhythm is measure,
But this Music of Nature is related to pleasure.

If one starts off by sharing, by sharing of timing,
Give a 'key' to the new world, a stairway worth
climbing.
If the 'key' is returned you escape to a new life,
For your stress is reduced as you part with the old
strife.

Such a 'key' now being used by 'ecofaces' of
places,
Such a door opens wide and the new world
embraces.
It's a world where a message is free from a
symbol,
It's a life where the actors are giving and humble.

For each message is based not on signals and
signing,
But on synchronization, on lateness and timing,
It is not what you do, what you touch, what you're
saying,
It's all WHEN you do deeds, a foundation worth
laying.

Then you're back to the former when hunger
encroaches,
You feel stress, and see signals as nurture
approaches,
But your message is signs, there's no feel for the
timing,
And your feelings are words, you are nickel and
'diming'.

So you move to the good world of synchronization,
And send 'key' to get 'key', unlock door to new
nation,
Where the feelings are sent by the times of your
motion,
And the bliss is as great as the moon on calm
ocean.

So the Music of Nature is rhythm of TURNING,
From the signals of old world to timing and
learning,
From the timing of new world to old world and
straining,
From the straining of old world to loss of such
paining.

Yes the Music of Nature is playing all about you,
And some creatures of Nature prefer now to shout
thro',
To proclaim that a world peace is destined to
transpire,
If we take from the new world and give to the
human choir.

PETER BEAMISH was born in Toronto and graduated in Engineering Physics from the University of Toronto in 1962. He then received a Canadian Trust Fund Scholarship to attend the Massachusetts Institute of Technology for a program in Geophysical Oceanography, and after graduating with an S.M. degree in 1964, he taught physics and mathematics at Phillips Academy in Andover Massachusetts for two years. In 1969 he completed a Ph.D in Bioacoustical Oceanography at the University of British Columbia under Dr. Robert W. Stewart.

Dr. Beamish studied whales, dolphins and porpoises at the Bedford Institute of Oceanography in Dartmouth, Nova Scotia for 10 years before starting his own laboratory, Ceta-Research Inc.

(Ceta is short for Cetacean or whale, dolphin and porpoise), in Trinity (Trinity Bay), Newfoundland. He has travelled extensively. Previous publications include descriptions of underwater sounds and behaviours made by many of the baleen whales, and the books "Dancing With Whales" (Creative ISBN 1-895387-28-0), and "TIME" beyond "conventional time" (Rodway's ISBN 0-9689955-2-7).